LIBERTY BONDS AND BAYONETS

A Marguerite Martyn Book

GEORGE GARRIGUES

CITY DESK PUBLISHING

City Desk Publishing
480 Morro Avenue, Suite E
Morro Bay, CA 93442

For permission requests, write to the publisher at the address above.

For more information, go to www.CityDeskPublishing.com.

Printed in the United States of America

1. Table of Contents

Introduction

THIS IS THE SECOND IN a series of books which revive the work of a notable American journalist whose illustrations and beguiling newspaper articles captivated the public for almost four decades of the Twentieth Century.

You may have met her in the first book, "Marguerite Martyn, America's Forgotten Journalist." Her husband was Clair Kenamore.

In this book, you will see these two, as unprepared for war as the rest of the country, confronting a world that was being reshaped all around them. They stepped up and did their jobs, like other Americans.

The text is in a form which may be new to you.

The ongoing story is taken directly from the microfilmed files of a variety of American newspapers. That story is set in Roman type. My comments are in italics.

You follow the yarn as everybody did in most of the 20th Century — in day-by-day newspaper accounts. There was no internet, no television, no radio. You bought the paper from a newsboy on the street or picked it up off your front porch where another lad had thrown it. Sometimes it came in the mail, a day late.

I tell you when and where the articles were printed. Unless otherwise noted, they are all from the files of the St. Louis Post-Dispatch, where Martyn worked from 1905 to 1941 and Kenamore from 1907 to around 1930. I've used their bylines only for the first in a series of reports by either of them. If you wonder who wrote any given story, check to see whose byline was on the previous article.

I've edited the text to make everything more understandable. The dates will help you look up the originals, which you should use for your high school term paper or doctoral dissertation or Wikipedia contribution, not this adaptation.

I invite you to linger over Marguerite Martyn's detailed (and often whimsical) drawings as her fans did. Take some time to appraise each character in a crowd scene or each drape, fold, and flounce that she used to lovingly signalize the women's fashions of her time.

Think how much has changed, and how much has remained the same.

Now, step back more than a hundred years. You are going to a movie.

1. Light Dawns
in Darkened Theaters

IN 1914, THE CHICAGO TRIBUNE sent photographer Edwin F. Weigle to film the German onslaught in Belgium, the beginning of the Great War in Europe, which we now call World War One.

For the first time, shocked American audiences could sit in the confines of darkened movie theaters, with only the whir of a movie projector behind them, and see how civilization had broken down in Europe. (Photo, Chicago Tribune, October 27, 1914.)

"Tribune" War Photographer
in Midst of Bursting Shells.

Under the simple title, "Belgian War Pictures," *the Weigle film caused a sensation.*

WAR IN ALL ITS NAKED UGLINESS

By Grace Kingsley, *Los Angeles Times*

December 11, 1914. Do you want to know what war really means? To view a vivid cross-section in all its naked ugliness, go to the Trinity Auditorium and watch the Belgian war pictures.

They are not pretty, these films. They do not always stir the imagination. The photography is not perfect, as in the made-to-order battles of moving pictures. But the scenes are the real drama of war, with all the heartbreak and pity and sordid misery of it; aye, even with occasional grim humor, for humanity smiles bravely whenever it can.

We first visit Belgian cities in peacetime, a fit introduction to the more intimate glimpses later on —

The cities Alost and Aerschot and Antwerp burning, and the crowds fleeing, in pitiful, halting groups, looking back in fear, or forward in stolid despair, or at each other in a hopeless groping for help. Sometimes there is a brave smile, sometimes laughter, from carefree little children or from adults under stress.

Old men and women totter with packs on their backs, cripples hobble on crutches, mothers carry babes in arms; all hurry away from their devastated homes.

Moment of Sacrifice

Pitiful incidents: The dying Belgian soldier, appalling, repulsive, not noble at all, big only in this moment of his supreme sacrifice; a sheep struggling in the meshes of the cruel electric-charged barbed wire stretched over a roadway; a peasant viewing the remains of his farm; a battle where the photographer is incredibly within fifty feet of the fighting — wounded soldiers being hurried away, replaced in a ghastly, business-like manner by others running to take their places on the front line.

The burning of a house, and the flight of its residents; the mother and daughter of a prominent Belgian physician, begging their way to neutral Holland; women in rich furs walking beside the hoi-polloi in the awful democracy of national disaster.

Sherman has said what he thought of war *(William Tecumseh, in 1879, considering it "hell"),* but you perhaps have envisioned it as orderly marching regiments and jaunty cavalry, directed by impressive officers on prancing horses, pawns in a big game of fortune.

If you want to understand war as it really is — not as a glittering spectacle, but in its human aspect, with all its misery, its oppression of the weak, and exploitation by the strong, you must see the Belgian war pictures.

In Muncie, Indiana, the Columbia and Star theaters scheduled viewings every hour from 1:30 to 9:30 for three days, Friday through Sunday, at 25 cents per ticket, with half going to a local poverty charity and half to Belgian war relief. "It isn't a play, it is the real thing," *warned the Muncie Star-Press (December 12, 1914, including the graphic following).*

A SCENE YOU CAN RECOGNIZE ON SCREEN

ONE of the most pitiable phases of the invasion of Belgium was the hardship worked on women and children. While their husbands and sweethearts were in the trenches, they were compelled to flee before the merciless fire of the enemy, taking what few household goods they could carry. To many of them it was a parting that was forever, for their loved ones lay dead before they themselves had reached a zone of safety. The accompanying photograph shows two Belgian women and a boy fleeing from Antwerp. It is just one of the hundreds of pathetic scenes from the film being shown at the Columbia Theater for the Belgian government and the Winter Relief Association.

BELGIAN WOMEN FLEEING BEFORE FOE.

In Richmond, Indiana, the Palladium-Item reported —

TRIBUNE FILMS DESCRIBE TERROR

December 3, 1914. It is a strange sensation to sit safely in a quiet, darkened theater and to watch the hideous drama of war enacted before one's eyes — to see the wreckage and ruin and desolation — to see men go coolly about the business of slaying one another, and to know that what you are watching is actual WAR.

This is what hundreds of people are doing daily where the Belgian war pictures are being shown.

There is no posing, no striving after effect; that uniformed lad smiles boyishly into the camera and into your very eyes. He salutes as he marches past. That boy is playing no part; he is going to the defense of his country and all too probably is now among the "missing" or lies dead in a Belgian grave.

The utter terror and bewilderment on the face of that Belgian woman as she flees before the oncoming Germans, clasping her baby to her breast and glancing back over her shoulder in fright at the bursting shells — this is no clever piece of acting.

The pictures have a strange effect. The people who crowd the theater talk in whispers. It is not uncommon to see a woman wipe tears from her eyes, or to hear an expression of sympathy burst from a man's lips.

Marguerite Martyn went to see it. She found that —

CAMERAS CONVEY NEW IDEA OF WAR

By Marguerite Martyn

December 16, 1914. Painters and illustrators strive with all the art they can master to bring home the horrors and, they say, the glories of war. But, accepting the truth of the camera, as we must, the scenes spread before us at the Garrick Theater via the Vitagraph every day this week give the lie to the great paintings of battle as mere stagey and theatrical claptrap.

No, in these "movies," we see nothing of brilliant military accomplishment, of tactical and orderly grouping, of flashing steel and charging steeds, of the smoke, blood and carnage which painters resort to in order to impress us.

By these moving pictures, we become intimately acquainted with that smiling soldier some fifty yards in front of us. He waves a greeting in our direction, then he tosses away his cigarette. He picks up a bullet somebody has dropped on the ground, loads his rifle, and runs up a little embankment under some willow trees. We see him place the gun to his shoulder and be swallowed up in a group of soldiers, with their backs to us, who are firing at the enemy.

Suddenly, we see two arms flung in the air. A figure falls. Men wearing white armbands with dark crosses rush forward and seize him, then drag him down the slope. Oh, surely it can't be the chap we knew and liked so well just a moment ago! We've no time to identify him, for another, and yet another, episode follows, equally poignant.

And there are the motionless pictures equally telling. There is a quiet little street in a Belgian village called Alost. No sign of life anywhere. Five or six crumpled bodies of men not in uniform mutely testify to how life, business, homes, and even the dear remains of the dead are abandoned in this street. *(December 10, 1914. The photo shows a street battle in Alost. The soldier at bottom left has been shot; he may be dead. There is a body also next to the truck.)*

A boat goes by on a broad expanse of water, out of which the peaks and chimneys of houses protrude, treetops jut. A seething mass of people crowd it. This might be the river Styx, so many signs of death and desolation abound.

Perhaps, we hope, our friend of the cigarette was simply wounded, and not killed. We are transported half a mile from the firing line, the caption tells us. There is a building on a street corner. Double- and triple-decked motor ambulances arrive. Stretchers bearing indefinable shapes are handed in, ten or twelve to each ambulance.

Oh, joy! Some few hands reach out from those shapes, and we see them waving, albeit feebly, to the cheers of bystanders.

Magically we are elevated to a tower in a city where we may look around at miles of buildings spread about on all sides. It is as broad as St. Louis, but how tightly packed, how closely wedged! It is Antwerp, a thriving, bustling center of industry, culture, and art when St. Louis was not even dreamed of. Then we see a veritable bridge of boats, placed next to each other over a single night — and a footbridge laid upon them — so that hordes of people and animals can escape in fright from a burning city into neutral Holland just beyond. *(Photo following.)*

In an instant we have a clearer view of the fleeing populace. Three hundred thousand people, I think the legend read, escaped over this one bridge.

Women can but run away in times of war, but it is gratifying to see how briskly, how pluckily the oldest and the youngest of them step along. A babe is in her mother's arms, another clings to her skirts, and the older ones, wooden-shoe-shod, trudge beside them. Bundles are on heads. *(Similar photo follows; a Belgian woman pushes two children.)*

On we are whisked to the outskirts of a village where a most effective — it would seem — defensive measure has been taken. An ancient roadway has been torn up and piled high with its own paving blocks, and then strewn with a network of electrified wire.

In the midst of another broad sea of water *(Belgian defenders have flooded the fields to impede the German advance),* we come upon what had been somebody's laundry yard. A maid has waded out to gather the wash. She drops some into the flood, but smiles and seems to blame her embarrassment on the photographer.

• • •

As vividly and intimately as we have been taken into these scenes, we have a distinct sense of safety and security in our own home afterward, with our own woes shrunken so much we are now ashamed to own them.

While in this mood it is well to remember that there is suffering nearer to hand. It is consoling to feel that in viewing this motion picture — which is being shown from 11 a.m. to 11 p.m. — you have contributed 25 cents toward the relief of poverty and misery, that amount going to swell the *Post-Dispatch* Christmas Fund to assure that the poor and ill and old or helpless of all this city have at least one bright, big day.

• • •

In the next chapter, we see what life was like in the Midwestern United States around the same time.

2. Peacetime

ON HOT DAYS IN CITIES over the country, kids could cool off by running after the water wagons that cleaned the cobblestone streets of horse droppings and other crap, like cigar butts, empty beer bottles, and yesterday's newspaper.

Marguerite Martyn did that drawing for the St. Louis Post-Dispatch's June 11, 1914, issue.

In my previous book, I dubbed Martyn as "America's Forgotten Journalist." But she wasn't forgotten in St. Louis, which was the fourth-largest city in the nation, after New York, Chicago, and Philadelphia.

That's because she was the only reporter in the country who was able to do an interview and at the same sketch the person she was talking to. Martyn was an arts graduate of the city's Washington University. She was a quiet, constant, and dedicated worker who had begun on the P-D in 1905 at the age of twenty-six as a simple editorial artist.

By 1913, she was a phenomenon: her drawings were featured about twice a week, most often with her bylined article to go with them. Or with a simple caption, which she wrote herself.

• • •

In March of that year she went up to D.C. to cover the inauguration of President Woodrow Wilson. She drew, below from left to right, Wilson's daughter Margaret and his wife, Ellen, then daughter Eleanor, Lois Irene Kimsey Marshal (wife of Thomas R. Marshal, the incoming Vice-President), and, standing, Wilson daughter Jessie. The woman with her back to us was Jessie Woodrow Bones, a Wilson friend from childhood (March 9, 1913).

Marguerite Martyn at the Inauguration of President Wilson

Martyn wrote a fulsome yarn about the event. She poked fun at a Missouri colonel waiting to take part in the five-hour inauguration parade; his horse ate the tassel dangling from the officer's sword. Trying to protect it, the poor guy got the blade jammed all crooked into the scabbard, so, Martyn observed— "They got a tinsmith to extract it later."

In 1913, her articles and sketches were published on seventy-seven days of that year. She probably was at the job six days a week, or at least five-and-a-half — the normal schedule for many workers. She interviewed and sketched mostly women; it was what she was paid to do. She covered "the female angle."

• • •

In May and June of 1913. Martyn took time off for a marriage and honeymoon with Clair Kenamore, the newspaper's telegraph editor. Well, they both took time off.

MARGUERITE MARTYN
WEDS NEWSPAPERMAN

Clair Kenamore Is the Bridegroom

KAJIWARA PORTRAIT
MRS. CLAIR KENAMORE.

"MARGUERITE MARTYN WEDS NEWSPAPERMAN" was how a rival newspaper, the St. Louis Star, splashed the wedding, with a two-column photo by Takuma Kajiwara, one of the country's leading portraitists. Wasn't it nice of the Star to be so generous with its space?

One sports columnist for the Star, in fact, got so carried away by this local journalistic coupling that he wrote:

OVER THE OLD HOME PLATE

By Billy Murray

May 19, 1913. Congratulations, Clair Kenamore and best wishes, Mrs. Marguerite Martyn Kenamore! We drink the Elks *[lodge]* toast to you, with our brothers. "Swing the goblet aloft, to the lips let it fall, then bend you the knee to address her, and drink, gentle sirs, to the queen of them all — Mrs. Clair Kenamore. God bless her!"

• • •

Bachelor Clair Kenamore had joined the Post-Dispatch six years earlier, in October 1907, so it wasn't a whirlwind romance.

At age thirty-two, he had walked through the doorway of the Post-Dispatch's editorial department and perhaps he eyed a tall, slender, dark-eyed twenty-eight-year-old woman with reddish-brown hair working quietly at an artist's easel. For a long time, they were teammates in the bustling news room, filled with clattering typewriters, chattering telegraph instruments and the smoke of half a dozen cigarettes. One thing led to another, and —

The ceremony took place on May 17, 1913, in the home of the bride's mother (Florence), on Lake Avenue in suburban Webster Groves, at 5 in the evening, with Episcopal rector Courtney Jones officiating. The newlyweds went to the Ozarks for their honeymoon.

• • •

Clair Kenamore (above, in 1918) had come to St. Louisans' notice in an 1897 Post-Dispatch interview, when he was a recent college grad on the way to Canada's Yukon to make his fortune in the gold fields. That didn't work out, so two years later he was private secretary to Congressman Edward Robb in Missouri's Perry County. (Kenamore's father was a Shannon County Democrat, so maybe there was some pull involved.)

His first newspaper job was on the St. Louis Republic, and then he went to the Chicago Tribune, where in 1905 he played golf for the Trib's team in the annual newspapermen's championship (handicap: thirty).

In 1906 the Chicago newspaper bylined him as "Claire" (a woman's name) for a story headlined "Farmer's Wife As Shopper Gets Her Money's Worth." (Just an error, or was the Trib trying to put something over on the reader? Or was it an inside joke?)

He moved to the Post-Dispatch in 1907, as telegraph editor, which meant he handled all the stories that came in by wire from outside the St. Louis area.

In 1912 Kenamore had a tiny role in an amateur production of John Galsworthy's recent play "The Pigeon." Another hobby: billiards. So he was a well-rounded guy.

• • •

The newlyweds came back from their honeymoon in July.

Afterward, Martyn:

- *Walked a picket line with women telephone strikers. She wrote that they had* "relentless purpose and not the faintest intimation of surrender." *(July 6.)*
- *Interviewed a fourteen-year-old girl suspected of causing seven typhoid deaths in an orphanage.* "Can anyone imagine a more lonely, more unwelcome, more desolate little figure?" *(August 6.)*
- *Went to Chicago to investigate the mysterious first use of women as police officers in that city and walked a beat with one of them. (September 14 and 28.)*
- *Sketched the fall styles as a circus. Why? Because women jumped through hoops and did tricks in the name of the latest fashion. (Next drawing, September 18.)*

- *Went to a lecture on beauty and heard the female speaker declare,* "Every woman's chief occupation in life is to GET a husband or to KEEP one." *She was flummoxed by that nonsense, observing wryly in her article that marriage was apparently* "to be a life's job, with long hours and no holidays." *(December 7.)*

. . .

Her last assignment of the year: Tell readers how Society People spent the final night of 1913. And draw pictures of them doing it.

WELCOME TO 1914!

By Marguerite Martyn

January 1, 1914 *(printed January 4)*. My memory of last evening is none too good, and I don't trust it. So I have notes and sketches made on the spot; right in the midst of the dozen hotels and restaurants where reservations had been sold out for a fortnight.

Twelfth Street was deserted at 10 p.m. One lone woman was bumping along from side to side across an empty street, a cow bell attached to her heels, and she was crooning a lullaby. Everybody who was anybody was inside; the doormen and the waiting chauffeurs were standing idly, smoking cigarettes, and some were sleeping.

At 11 o'clock, a theater opened its doors wide, and out poured the evening's audience, hastening in every direction in the last hour before the end of 1913.

I hurried to an improvised dressing room on the second floor of the Jefferson Hotel, where, amid a crushing crowd and, almost asphyxiated by the thick odor of perfumes, I struggled to hand over my wrap in exchange for a cardboard coat check.

I threaded my way through a dizzy maze of tables and chairs blocking hallways and parlors before I could reach my host's table in the middle of the big dining hall. Somebody helped me stand on a chair where I could gaze over what seemed to be a field of multicolored tulips or peonies. No! They were actually the thousand and one varieties of paper caps atop an equal number of celebrating St. Louisans' heads.

I could easily pick out the newcomers from the old-timers by their eyes filled with wonder at the display. *(Next image.)*

THE DÉBUTANTE AND THE GIRL WHO WAS THERE LAST YEAR

The thousand people all talked at once — punctuated by the boom-boom bass drum from the orchestra and the staccato pop-pop of champagne corks — and everybody tested their whistles, cowbells, tin horns, and frying pans beaten with spoons. Waiters hurried and scurried, stampeded, and stumbled.

Then, suddenly! It was close to the hour of midnight. Out went the lights. All scrambled to their feet. Into the air were raised the wine glasses. The room quieted, except for a giggle or two. *(Next image.)* I spied a waiter with an opened champagne bottle attached to his lips, and he drank deep. (And I had just been feeling sorry for him because he had to work on New Year's Eve!)

The first stroke of midnight pealed from the new Cathedral across the street, and a gamut of chimes followed.

Within our own hotel cloister, electric lights sprang forth to spell out "1914" and "A Happy New Year."

Now, the clinking of a thousand glasses and — everybody kissing somebody, usually the person nearest at hand (one's wife or husband, preferably). Failing the lack of a suitable female close by, many a strong man was seen trying to implant a New Year's salute on the cheek of an objecting brother. *(Next drawing, at top right.)*

The celebrating boom of artillery rumbled from distant guns.

Finally, the kissing was over (almost), sparklers were lit, confetti was showered, and gay serpentines were thrust over the happy throng. (The people at the next table over, however, kept up a kind of kissing contest for the rest of the morning.)

A nine-course dinner had to be put away, to keep up with the wine being drunk. Then, visiting time! You knew nobody in the room? No matter; you just moved from table to table until you found somebody interesting.

A dozen celebrants approached a pink dumpling of a woman seated near me.

"You have such a kind face, I want to wish you a Happy New Year!" they would say. *(Next sketch.)* She got quite used to it.

In a wee corner where the crowd was not wedged too tightly, people were trying to dance. Some men stood around and speculated on harem trouserettes and a lost shoulder strap. The women talked about the new show with Julian Eltinge. *(The famed female impersonator, one of the country's highest-earning actors, was starring in "The Fascinating Widow" at the Olympic Theater.)*

As the morning wore down, visitors flowed into the Jefferson from other restaurants, and the Jefferson people streamed out to have a look at what was going on at McTague's, Lippe's, the American, Planters, Faust's, and everywhere else. *(All of them places to eat and drink.)*

If you had the time, you could also take in Cicardi's and Cafferata's in the West End and even the Racquet Club if you were lucky enough to have friends there. Then you'd return to the Jefferson for breakfast at 6:30.

Some people should have left their wife or husband at home. Sitting around while others are dancing is a good way to head for the divorce court. *(Next sketch.)*

Some Husbands and Wives Might Better Have Been Left Home

But one man found that a wife was a handy thing to have with him as the garish dawn broke through the windows and coffee somehow began to taste better than the finest wine. This grandfather leaned upon the left arm of his missuz as they departed, veering off to the side with every other step. Then he'd balance safely to the right, and express his gratitude with a kiss on her bare, white shoulder. Every alternate step he hit the same spot precisely. Wonderful steadiness of aim!

She flagged down a taxi for both of them. *(Next image.)*

• • •

So 1914 began with great celebration.

3. War as Seen From Over Here

WE JUMP AHEAD SEVEN MONTHS. On Saturday, August 1, 1914, Germany declared war on Russia. Eventually the whole world was caught up in the terror.

Three days later, Kaiser Wilhelm's big army plunged through little Belgium on its planned drive to Paris. By the end of the month, the Belgian army (what there was of it) was almost gone. (Front page, August 27, 1914.)

ST. LOUIS POST-DISPATCH NIGHT EDITION

GERMANS TAKE LONGWY, ALL NAMUR FORTS

Fallen Cities Were in Path of German Center—France Also Suffers on Extreme Ends

Russian Advance Into Prussia Rapidly Nears Posen, 150 Miles From Berlin

RUSSIAN ADVANCE ALARMS BERLIN AND RICH DESERT CITY

NEW WAR MINISTER PREPARING PARIS FOR A POSSIBLE SIEGE

BRITISH CRUISER SINKS GERMAN WARSHIP OFF WEST AFRICAN COAST

In St. Louis, the Star interviewed "merchants along Washington Avenue" *who felt that the war would be good for business. Somebody suggested that, to boost the war effort, city parks should be turned over for national defense, so troops could drill there.*

Most folks looked at war in Europe with dread and anguish. A Post-Dispatch editorial opined:

PEACE SHOULD BE OUR GOAL

August 22, 1914. To the suggestion of an ex-member of the National Guard of Missouri that we build barracks in the city parks and use the greenswards and playgrounds for drill grounds, we recommend the statement of the International Conference of the Church Peace Union: "Peace is not to be secured by preparations for war."

In one of history's oddest quirks, this little-remembered peace meeting had begun in Kostanz, Germany, on August 2. The next day, August 3, Germany invaded Belgium; Britain and Germany declared war on each other on August 4. The church conference recessed, and a train carrying peace delegates through Germany to the coast was guaranteed safe passage by Kaiser Wilhelm. The conference reopened in London on August 5, even though by that time the German Army was in its second day of a murderous swath through Belgium.

The Post-Dispatch editorial continued:

Events have shown that it is the momentum of half a century of war preparations that has at last brought about the catastrophe in Europe.

The wisest policy at this time is to think peace, speak peace, and live peace. To turn the parks in the great cities into armed encampments would simply excite the public mind and develop a war spirit. As long as we keep calm, as the President has advised, no nation can or will attack us. It was mobilization that lit the flames in Europe.

Not many St. Louisans wanted the city's Forest Park, as Martyn had sketched it on the opposite page, September 10, 1913, to be turned into an "armed encampment."

The P-D continued:

Reasonable provision for the militia is wise and proper, but we need our parks for their present uses. Only the direst necessity will induce the people to give them up to the military.

You will hear of the "militia," or National Guard, again, starting with Chapter Five. And Forest Park remained tranquil for civilians.

• • •

In the first month of the war, Marguerite Martyn drew a telling cartoon that summed up a stereotype: Women didn't give a fig about the news of the day; that kind of stuff was the province of men.

I am sure she herself did not believe this, but her newspaper persona was often tough on the flighty women whose emptiness drove her crazy. She pictured a vapid woman lounging the morning away

Just a Hint to Those Who Appear to Think There Is No Interest in Anything Save War

Drawn for the Post-Dispatch by Marguerite Martyn.

THE LANGUID ONE: Not a bit of news in the news papers, nowadays—not a bit—except that the prices on hair nets and silk stockings are positively prohibitive!

while three men gathered avidly about a newspaper on the other side of the veranda, sucking up the war news.

THE LANGUID ONE

By Marguerite Martyn

August 27, 1914. Not a bit of news in the newspapers nowadays — not a bit — except that the prices on hair nets and silk stockings are positively prohibitive!

• • •

The next day, the Post-Dispatch informed its readers on Page One that

GERMANS KILLED TWO NURSES

August 28, 1914. *(No byline.)* France has submitted to the United States and neutral governments a sworn statement that after an engagement at Moncel, France, a German officer fired on three Red Cross nurses, killing two and wounding a third.

Whether that dispatch from Washington was true or not (in wartime it didn't seem to matter), on an inside page that very day, Martyn had thought it would be cute to illustrate what Wartime Fashion might have in store for women shoppers. Bad karma. In the center, there is a Red Cross nurse (August 28, 1914).

On the left, a society woman gives her time to a community kitchen (that glittering diamond bauble draped from her neck is dangerously close to a soaking in the soup). Next is a woman rigged up in military attire. Then the nurse.

Fourth, there is, well, perhaps a motorcycle driver with goggles (or maybe she's an aviatrix) and, at far right, a journalist dressed like Sherlock Holmes. Half seated on a motor bike, she looks through field

Autumnal Fashion Forecast . . Drawn for the Post-Dispatch By MARGUERITE MARTYN

PUBLIC SOUP KITC

If, as usual, we look to Paris for styles, this fall, these probably will be the most popular models.

glasses and makes notes on the pad at her knee. The envelope in her pocket reads "Dispatch to Le Temps," *the leading and most serious French newspaper.*

(American women writers such as Mary Roberts Rinehart, Corra Mae Harris, Inez Milholland, and Alice Rohe did cover the war, but none of them dressed in checkerboard suits with matching caps.)

• • •

Wars had been fought in 1912 and 1913 in Bulgaria, Greece, Russia, the Ottoman Empire (Turkey), Romania, Montenegro, and Serbia, in various combinations. In the U.S., unless you were Bulgarian, Greek, Russian, Turkish, Romanian, et cetera, you didn't pay much attention.

Martyn took note of who was quarreling with whom. For years she had sketched St. Louis women wearing the latest designs from Paris, but now there was a war on! So three weeks after Woodrow Wilson's proclamation of American neutrality, we see —

NEUTRALITY LAW VIOLATED BY DAME FASHION

By Marguerite Martyn

August 30, 1914. The Balkan Wars provided fashion designers with enough ideas to satisfy fickle fancy through several seasons. We've scarcely recovered from blouses with Magyar *(Hungarian)* sleeves, Bulgarian embroideries, Turkish sashes, and harem skirts.

Strange is the caprice of Fate that will let one nation's sorrows serve as a means to gladden the raiment of another country's women! But the present wars are proving a windfall to fashion designers. A national milliners' group decided in Chicago last week that all fashions shall be of strongly military character.

Yes, the Retail Milliners Association had indeed proposed new styles based on military themes. (Anything to sell merchandise.)

Martyn offered these tongue-in-cheek fashion ideas for milliners to copy:

- A helmet-shaped bonnet to be known as the Kaiser Wilhelm.
- A long, low, visored toque *(small, brimless, close-fitting hat)* to be called an "uhlan" *(a lancer in Central and Eastern European cavalry).*
- A three-cornered hat trimmed in military fashion with binding and rosette of grosgrain ribbon or gold braid, answering to the name of Poincaré *(the President of France).*

- A bonnet flaunting coque feathers named after Earl Kitchener. *(More bad karma. He was the British secretary of state for war who was killed the next year, 1915, when a ship was sunk by a German mine.)*

There are cossack turbans, Belgian toques and Highlander bonnets, and there are cute little Servian caps *(the old way of spelling Serbian),* and tomorrow someone might design a turban shaped like a submarine to be known as the "Rear Admiral Beatty." *(Frank E. Beatty commanded a division of the U.S. Atlantic fleet.)*

Hats to Reflect Nationalities?

But suppose women should take all this seriously and find in the new hats a way of expressing our own national sentiments?

Then, my! What battles might ensue in the bonnet shops when the new fall hats arrive in our heterogenous St. Louis?

Imagine an insistent saleswoman trying to persuade a sympathizer of the Russian Czar that a hat named "the Kaiserin" *(for the Kaiser's wife),* looks just darling atop her coiffure! *(Not likely. The Russians and the Germans were killing each other in Eastern Europe.)*

Women would snub each other over their fashion choices, as on the opposite page.

Suppose a small Servian woman is wearing a Russian blouse, a Poincaré hat, and a pair of English walking boots — and she should meet a large lady in a German military cape and helmet — what would happen? ("Now what is the matter with Frieda Schnitzelwurst? Passed right by me and didn't speak!*")*

Friends and sister style sufferers, do you think we should risk any international complications by donning these proposed fall styles?

But there is a fashion solution. The milliners' convention also brought forth an all-American hat called "The Woodrow," after President Wilson.

I have dared to draw my own conception of what a truly Democratic hat should be. Of course it will have to have long ears *(for the Democratic donkey)*.

Better to wear "The Woodrow" than risk any battle with another woman by having a foreign bonnet on your head.

A COMPROMISE IS SUGGESTED, "THE WOODROW"

• • •

The light-hearted mood did not last. People were being killed overseas (next chapter).

4. Belgium

ONLY A SLIVER OF BELGIUM remained free. German troops occupied the town of Louvain, where they burned and looted and executed hundreds of civilians. The destruction and murders there became the symbol of the brutal nature of the German war machine.

• • •

In May, a Catholic nun named Marie Ignatius was in St. Louis, a refugee from the terror. Martyn went to interview her.

A TOWN OF SILENCE AND DARKNESS

By Marguerite Martyn

May 23, 1915. The population has returned to Louvain, but it is to fulfill the desire once expressed by German Chancellor Otto von Bismarck that in their next war the German armies would leave the civilians "nothing save eyes with which to weep."

The speaker was Sister Mary Ignatius O'Kavanagh of the Daughters of Mary Convent, Louvain, Belgium, who — with her traveling companion, Sister Mary Augusta — is a guest of the Sisters of Mercy Home, 23rd and Locust streets.

These good women are seeking aid in America for the men and women of Belgium, many of them destitute and with no means of making a living.

"Yes, the people have returned to Louvain," Sister Mary Ignatius continued. "They were permitted to do so six weeks after that terrible Tuesday, August 25, when we were herded together and forced to walk to Tirlemont, five days away, with a German escort prodding us on.

"The nuns of my order were allowed one-and-one-half minutes to evacuate our convent. We had not time to go upstairs and collect linen — we were not given time to gather a little food or anything we might need for a journey."

In Martyn's drawing two nuns work at looms, and at right three young women admire the delicacy of another sister's stitching. The nun on the left holds up her hand to still the chatter, while a phalanx of flag-waving intruders rush menacingly toward them. Martyn has drawn the enemy as javelin-wielding marauders of olden times, not as modern German soldiers. A cat raises its fur.

"WE WERE ALLOWED ONE MINUTE AND A HALF IN WHICH TO EVACUATE OUR CONVENT" SR. M. IGNATIUS O'KAVANAGH.

"Houses were fired simultaneously all over the city by naphtha-filled grenades. There were no fire escapes such as you have here, and the invaders had intercepted all fire-fighting apparatus. As we looked back over our shoulders, fleeing on, we wondered how many of the townspeople had not had time to escape from the upper stories of those burning buildings.

"For six weeks we lived in sheds at Tirlemont — those who hadn't been driven still farther on into Denmark *(a neutral country, but some five hundred miles from Belgium),* and those who hadn't died of heat and exhaustion on that terrible march. And every day at sunrise two men were selected from among their families by the German guards, taken out and shot.

"At last there came a general of high rank. He asked what we had done to provoke such treatment, and there were tears in his eyes as he beheld our plight. He promised to send us back to Louvain, and two days later trains arrived.

"Many had to stand all the way, and the journey, which in ordinary times took but a few hours *(to travel a dozen miles),* was two-and-a-half days long. A woman gave birth to a child on the train, and the others had to huddle closer to allow her a space on the floor."

"Did the people want to return?" I asked. "Did they not dread going back among the enemy?"

"Few hesitated," she replied. "It is the same instinct that impelled the French nuns who were expelled from their country to settle just as close to the border of their native land as possible and hope always that they would return to France." *(During the French revolution, 1789-1799, many nuns fled to Belgium.)*

Sister Mary continued, "We returned, but Louvain was no more. The people went back to the places where their houses had been, cleared off the debris, and lived in the cellars, erecting some sort of roof for shelter against the coming winter.

"The history of Louvain down the centuries is always interesting. You have read much about the medieval landmarks, the great university, and the library, which was one of the world's wonders. I wish I could picture to you the life of the city before the Germans came. Belgium was an intensely commercial little country. Everyone was in business,

but they had a fine regard for recreation, and they made the most of their leisure hours.

"They are a musical people. With a population of fifty thousand, Louvain had more than ninety musical societies — the shoemakers' chorus, the carpenters' chorus, and so on. In the evening music and singing was everywhere.

"And now, all is silence. All is darkness. We are ordered to have all lights extinguished at 8 o'clock. Nobody must be upon the street. Our doors must be left unbolted. A guard with a lantern patrols every section. You cannot imagine how depressing the evenings are in Louvain. The awful hush — it is terrifying."

Irish Accent

Sister Mary Ignatius was born in Limerick, Ireland, and even after forty years in Belgium, she has a trace of a brogue. If there were the least opportunity for an Irish jest or a smile, she would not let it slip.

But I fancy it has been many months since she has really smiled. She does not speak with rancor or bitterness, but neither does she have the air of patient resignation that one might expect of a *réligieuse*.

Sister Mary went on:

"It is not prudent for me to antagonize the Germans while my people are still at their mercy. I hope America does not enter the war, if only to save the Belgians from the Germans' further vengeance. The American relief supplies would be cut off, and Belgians would be left to starve.

Belgians faced starvation under German occupation, but, finally, food started to arrive in a massive relief effort under the command of a wealthy mining engineer, Herbert Hoover, later the 31st President of the United States.

"And, too, I do not want to condemn wholesale the German people. Individually, the soldiers seemed to have no hard feelings toward us. I have heard them say repeatedly, 'It is a pity to destroy your beautiful little country.' I have seen them perform many individual acts of kindness.

"That nuns were ravished, priests persecuted, civilians fired upon, children maimed, that there were unspeakable cruelties everywhere — these things were done at the command of superior officers. We are sure of that, in order to terrify the people into subjection.

"And the German soldiers hate their officers. Why shouldn't they? They are disciplined as if they were criminals and never addressed by any word other than 'swine.'"

Nurses to the Wounded

What was her object in coming to America?

"It was to collect funds to build a home for impoverished women. You see, the four buildings of our convent are filled to overflowing with refugees.

"Before the war, we had a boarding school, where we taught the daughters of the well-to-do, but also trained women in domestic pursuits, to be self-supporting. We gave employment to hundreds, through our needlework and our extensive connections with lingerie dealers in Paris.

"It will be a long time before we can re-establish these industries upon the same scale" — she said this so sadly — "because in the meantime we have to house and feed the destitute. We have a large tract of land where we could erect a shelter, and it is with the hope of interesting some of your good women in this project that I am in St. Louis."

5. On the Border

*AND WHAT OF TELEGRAPH EDITOR Clair Kenamore? Weren't he
and Marguerite Martyn married back in Chapter Two? And didn't they
have a happy family life and lots of babies?*

About the happy family life, I can only surmise: Yes.

About the babies, I can tell you: No.

*I've plumbed the sources to find out more on these two. They left
no progeny to pass along family stories or anecdotes. She had two
unmarried brothers, and he had a nephew. What we have are simply
their words and pictures, a few public records, and what other people
wrote about them.*

*I know that he was on the Post-Dispatch's billiards team, which in
1911 played in the Scribes' Tournament against the Globe-Democrat,
the Times, the Star, and the Republic. Yes, there were that many St.
Louis newspapers in those days, plus some in foreign languages
and at least one that catered to African-Americans.*

• • •

*Kenamore, though a ruggedly big man, wasn't in very good health.
Over the winter of 1914-15, he went out to Arizona for some desert air.
He bought a Ford and registered it in Puma County. He got his name
in the Tucson papers a few times for social activities he attended.*

*He tried his hand at freelance writing. At least one article was
published in a string of newspapers — an odd little piece about golfing
on Guam, of all places.*

As well, I can envision him filling reams of paper with letters to his wife, sharing with her the details of his days and the sights and sounds of the Southwest.

News was happening quite close, he would have written. Just a few miles away, revolutionary war had broken out. Bandits (or el ejército del comandante Francisco Villa, to many Mexicans) were raiding American settlements in Arizona and Texas. The Associated Press and the New York World and a few other papers sent reporters.

Managing editor Oliver K. Bovard of the Post-Dispatch wanted his own man on the line. He was tired of relying on press agency reports or shared stories from other newspapers.

Perhaps he had to get permission from Joseph Pulitzer, who owned both the St. Louis Post-Dispatch and the bigger, more influential, New York World. Finally, Bovard chose Kenamore as the P-D's first foreign correspondent.

In the spring, the El Paso Times reported that:

ST. LOUIS NEWSMAN IS ON WAY HOME

May 29, 1915. Clair Kenamore of the *St. Louis Post-Dispatch* was an El Paso visitor yesterday, en route to St. Louis from Tucson, Arizona. Mr. Kenamore, who is one of the most prominent newspapermen in the Middle West, has been in Arizona for several months. He has fully regained his health and is ready to report for duty in St. Louis on Monday.

• • •

Bovard kept Kenamore around St. Louis for a while, but in the fall sent him off to the Mexican border. Kenamore at age forty did not speak much, if any, Spanish, so the only people he could easily interview were those who looked like him and his boss.

TROUBLES IN SOUTH TEXAS

By Clair Kenamore

November 7, 1915. The condition in Brownsville, Texas, is that the Americans, or "whites," are ready to destroy the "Mexicans," that both sides are armed, and that thoughtful men look to the next violence by Mexicans with the greatest apprehension.

The whites referred to all the Spanish-speaking people in Texas as "Mexicans," even though those born on the north side of the Rio Grande were U.S. citizens. Kenamore adopted this practice in his articles.

Racial prejudice, unscrupulous politics, religion, poverty, the hair-trigger methods of the Texas Rangers — they all get portions of the blame.

During the last three months, at least eleven thousand Mexicans have fled across the border. Crops are unharvested, cotton unpicked, and ground untilled because the laborers are gone.

A condition almost of terror prevails. Every Mexican's house is presumed to be an arsenal; lonely ranches are armed and fortified by their white owners.

Cavalry patrols clatter all night through Brownsville, and soldiers sleep beside their weapons because authorities believe a raid on the town is possible.

This is how it happens:

About five hundred bandits, trained and uniformed (after a fashion), with officers and a bugler, make sudden forays from Mexico across the Rio Grande, attack a detachment of soldiers or a ranch, and then disappear.

Who are they? Soldiers? Thieves? No.

Bandits Are Led by American

The leader is Luis de La Rosa, born in Brownsville, a United States citizen, a good Democrat and party worker, and formerly a deputy sheriff. It is probable that more than half his men are American citizens, but the right to vote for the President does not seem to make a Mexican any less loyal to Mexico.

Americans are willing to announce that they are better than Mexicans and that the latter "know their place." In Matamoros, across the river, it is vastly different. The whites see them as truculent, swaggering, and spoiling for trouble.

The ten thousand or so settlers here from our Northern States know nothing about the Mexican character, and the idle, dark men along the watercourses alarm them. Whenever a goat disappears, the newcomer rides into town and seeks a warrant, but the old-timers regard such thefts as a matter of course.

Up to last August 2, there had been no particular violence anywhere. On that day, though, a rancher named Scribner thought he had located the lair of some cattle thieves, so he obtained a detachment of soldiers and took them there.

Gunshots rang from the house as they approached, and Private McGuire was killed. Then the "stuff was off," as they say here, and there followed almost a reign of terror.

A posse of seventeen men under Sheriff Vann engaged in a fight with Mexicans at Paso Real, and three of the latter died. Two days later a man named Austin and his sons were killed at Lyford, possibly in retaliation.

Constant Flow of Violence

Since then — a constant flow of violence. In one instance, after a raid by bandits and a return visit by the Texas Rangers, the bodies of eleven Mexicans were found.

One Mexican woman told me that her family had just seated themselves at dinner when five Rangers rode up and took away her husband and three brothers. Five days later, her little boy found their bodies in the woods.

"The killing of an innocent Mexican makes from five to a dozen new bandits," a leading Brownsville businessman told me.

I protested to an officer at Fort Brown that the heavy guard and the obvious precautions seemed excessive and calculated to arouse resentment.

"We have a disloyal and hostile city just outside the fort," he replied. "Ninety percent of these citizens are Mexican, and maybe a third of them know the bandits, but we never get information from them. The troops are as much in a hostile community as we were in the days of Philippine reconstruction."

The United States had gained control of the Philippine Islands in the Spanish-American War of 1898.

Most Mexicans are of astounding ignorance. Their information comes from Mexican sources and is colored amazingly.

The difficulties of Colonel Blacksom and his officers are endless. They try to do police duty without police powers. But the mounted cavalry, each man with a rifle under his stirrup leather, do not reassure the Mexicans. These latter watch the horsemen from the sidewalks, unmoved or with sneers.

Attack Interrupts Concert

One night a band concert was held at the fort, to which the beauty and the chivalry of Brownsville had been invited. Fifty parked automobiles were filled with listeners, while the olive-skinned population mingled peaceably with soldiers all over the parade ground. It was a fine evening, and it looked like the Army's entertainment was gaining the friendship of many people.

Colonel Hatch was at his desk a hundred yards away when he received a telephone call from a sergeant at the nearby oil field. Bandits were making an attack!

Within ten minutes, the colonel had "borrowed" enough automobiles to load a company of soldiers and send them to the rescue.

The audience had to make their way to their homes on foot.

• • •

Kenamore went west to Nogales, Arizona, which is a twin city to Nogales, Sonora, in Mexico. They are separated by a wide street called Internacional, if you speak Spanish, or International, if you prefer English.

In the photograph, about 1898, a line of poles marks the border. The U.S. is on the left and Mexico is on the right.

Kenamore found Mexicans under Comandante Francisco (Pancho) Villa facing Americans under Colonel William H. Sage. Almost every man had a weapon pointed across the street.

BULLETS FLY ACROSS BORDER

November 27, 1915. For the first time in four years, American soldiers have fired across the line into Mexico without having to wire Washington for permission.

And for the first time in four years, Americans here are walking with shoulders thrown back and their cigars tilted upward at a prideful angle.

That's because the 12th Infantry shot up the Mexican side of the border. One American was killed and two wounded. Between sixty and eighty Mexicans are dead. Americans killed forty, and late-arriving Carranzistas slaughtered the wounded.

The Mexican forces (Villistas) were allied with rebel Pancho Villa, but late arrivals (Carranzistas) rode up under the command of Alvaro Obregón, who was fighting for Venustiano Carranza.

Yesterday morning, Colonel William H. Sage of the 12th Infantry knew trouble was coming, and his men had plenty of ammunition. The Villista men across the border were quarrelsome. For days they had been shouting insults at the Americans.

"Come out and fight, you cowards!" cried a particularly offensive Mexican as he fired his rifle. The instructions came down to "get" him the next time he shot. As he came into sight, the rifles of the Americans opened up, and the *bravo* went over.

General Rifle Fire

Some two hundred Mexicans began a general rifle fire, from housetops and ditches, from behind barrels and every other vantage point on their side. The American skirmish line rippled into fire from hill to hill.

Men who had been waiting for two years finally had their chance. Calmly and coolly, they proved their marksmanship. Every time a Mexican was toppled over, women ran out and gathered around him, greatly hindering the American soldiers.

For a quarter hour, the firing was sharp. Then American consul Frederick Simpich went to the barbed wire and called to the Mexican police chief.

"Why don't you make these fellows quit firing?" Simpich said. "A lot of people are being killed unnecessarily." *(Simpich later became a writer and was known as having done more articles for the National Geographic magazine than anybody else. You never know where a government job is going to lead you.)*

The chief said he would do his best, if the Americans would not shoot him. So word was passed along the line not to get the man with the white flag tied to his rifle.

Newcomers Make a Mistake

At this moment, a line of cavalry appeared on the hill to the west, on the Mexican side, immediately opening fire on the Americans. These were the advance of Obregón's column, and they had mistaken the Americans for Villa troops.

They soon found out otherwise and took cover. A major advanced from their line and held up a bleeding hand. "Why did you fire at us?" he called. "We are Carranzistas!"

This being verified and polite messages passed, that phase of the fight was closed.

By this time, the firing had almost ended, and many a Villista was taking the final step in the life of a *soldado*. He threw away his rifle and cartridge belt, tore the identifying band from his hat, and ran across the border into American territory.

• • •

Two hundred miles east, in Columbus, New Mexico —
In the early morning of March 9, 1916, when most everybody was
sleeping, Pancho Villa and what he called his "División del Norte"
swept across the border, burned down some buildings, killed U.S.
soldiers and civilians, then fled, with the U.S. 13th Cavalry Regiment in
pursuit. But the Americans had to pull back from Mexican soil.

Washington ordered a response.

State militias received commands to head for New Mexico. That
would free regular Army troops to follow Villa on what Americans
called a Punitive Expedition, to kill him or to catch him. Kenamore
was in Columbus and watched the regulars ride out. They were under
the command of a brigadier general named John J. Pershing.

TROOPS CROSS INTO MEXICO

March 20, 1916. Two troops of negro cavalry crossed the border yesterday to follow the American expedition. *(Photo following, from the Wilmington, Delaware, Evening Journal, April 6, 1916.)*

The troopers crossed the line without a quip, a smile, or jest. To all appearances, Mexican soil was exactly the same to them as American.

This casual diminishment of the black soldiers — in effect, that they should have been filled with jokes and didn't even know what country they were in — probably stemmed from Kenamore's impression of African-Americans only from what he saw on the vaudeville stage. Later, in Chapter 13, he has a different experience.

UNITED STATES NEGRO CAVALRY MARCHING THROUGH MEXICO.

Army aeroplanes flew far southward from the base at Columbus, circling the Mexican mountains which lie beyond Palomas, doubtless raising the spirits of the column of men and horses below them.

The wireless outfit the soldiers carry has been in uninterrupted communication with the station here in Columbus, where many other militia units stir about town and take practice marches into the country to keep in shape.

Major W.R. Sample, the commandant, keeps in touch with General John J. Pershing's forces by wireless. Communications of newspaper correspondents are censored.

Hurried preparations were begun today for a cantonment hospital, which is to coordinate with the field hospital and ambulances in Mexico and with the base terminal in El Paso.

A new set of officers appeared today, as is the case each morning. Veterinarians from as far as Brownsville are taking up the trail of the column. Medical men from many points are reporting, not so much for the expectation of heavy action as for preparedness.

An ambulance train of twenty vehicles, each drawn by four mules, went north on a drill and came back two hours later. The stretcher carriers rode with each ambulance, and medical officers in command were on horseback beside them. A train of nine mules swung through in a trot, each fully packed, and all following a sergeant in the lead.

There is a large demand by soldiers for baths. The only tub in town is kept busy day and night at a charge of 25 cents for each wash.

• • •

Word came that the militia troops still at home in Missouri were ready — the infantrymen fully equipped (except for shoes) and the cavalry eager to start out (except they had no horses). It's all there, in the June 22, 1916, issue of the Mexico, Missouri, Weekly Ledger. To be fair, the Washington, D.C., Times reported on June 29 that the equipment and steeds would be furnished at the border.

Clair Kenamore went back to Texas to take a look. He got there before most of the Missouri troops, so he advised them good-naturedly just what to expect after their lengthy train trip.

INTRODUCTION TO SOUTH TEXAS

June 30, 1916. The Missouri militia will have a widening experience in El Paso.

Militiamen will have ridden during a pleasant summer down the fertile valleys of Kansas, Oklahoma, and Texas into San Antonio, which they will come to know as "San Antone."

They will visit the Alamo and stand bare-headed in its gloom. After a while they will remark on the heat. A little later they will go out to Fort Sam Houston, pet the deer in the park, and wonder what is going on behind the gray walls that shelter General Frederick Funston and his staff. *(They were overseeing the hunt for Pancho Villa.)*

Our soldiers will observe with pleasure the spreading palm trees and with pain the price of a beer — 15 cents a bottle! *(As opposed to a nickel in St. Louis.)*

They will be called sharply to account by khaki-clad policemen *(instead of in blue, as back home),* and they will be reproved by streetcar conductors for sitting in the "Jim Crow" seats *(for blacks only, and there were none in the Missouri militia).*

Along about noon they will notice the heat, and between 2 and 4 they'll see the sidewalks are virtually empty, but they'll be full again between 5 and 8. Finally, somebody will inform them about the afternoon "siesta," one of the great blessings of the Southwest.

They will think that San Antonio is the hottest town in the country but they will be deceived, because they are going on to Laredo.

Kenamore also went on to Laredo, a hundred and fifty miles south of San Antone, right across the river from Mexico.

MISSOURI TROOPS LOOK GOOD

July 17, 1916. The Missouri militia is a mighty nifty organization and is (barring the Massachusetts contingent) the best body of state troops I have seen. *(You will read more about this spiffy, but essentially useless, Missouri unit at the beginning of Chapter 17.)*

The brigade is camped on the edge of Laredo in probably the worst spot for miles around. It was originally part onion field, part municipal dumping ground, and part the inclosure of the county pest houses *(where poor people with diseases were housed and sometimes treated)*. The soil forms thick dust in dry weather and deep, sticky mud in wet.

The four regiments of infantry are jammed into a space which would accommodate just one in comfort, and the ambulance corps is strung alongside the brigade headquarters. Brigadier-General Clark has fair space for the HQ, but only because he took more ground than the regulars had marked out for him.

The Missouri troops look good. Many are big, upstanding fellows, over six feet tall, and as they swing along the narrow Laredo streets, all eyes follow them. They seem of a different species from the poorly nourished Mexicans.

The troops are drilled at least four hours a day. They are better trained, no doubt, than many of the Missouri soldiers were at the beginning of the Civil War. They seem to need target practice, though most have fired the new Springfield rifles.

The Missourians are still in the heavy olive drab uniforms designed for winter, but an issue of the lighter cotton khakis is on the way, and if the quartermaster's department does not fall down, everybody will have a new suit this week.

Church services were held on a sizzling hot Sunday morning. The day was free from drill, the chow was good and so plentiful that the waiting row of Mexican children filled their baskets and buckets.

In the balmy evening following, the Second Regiment band played in the old Plaza downtown, where pepper, magnolia, oleander, and cottonwood trees fill the squares. Five hundred men lay on the fine, soft grass or sat on benches in the moonlight while they heard "Turkey in the Straw," the "Houn' Dawg" song and other tunes dear to the Missouri heart. About as many Mexicans were in the audience.

The band's mascot, whose name *is* Hound Dawg, was in the stand but refused to howl as expected.

6. Preparedness

ST. LOUIS WAS A PART of the Louisiana Purchase of 1803. We bought it from the French, and there were French people living there. The city was named for a French king — Louis IX, who had become a saint of the Catholic church in 1297. Can't get much more French than that.

In December 1915 the big social news in St. Louis was a three-day Fête de Charité on behalf of French war widows and orphans.

EVERYBODY IS FRENCH NOWADAYS

By Marguerite Martyn

December 15, 1915. Two distinct kinds of French people are helping to make the Fête de Charité a pronounced success — the old French and the new French.

- The old French: They are mostly Flannigans, Walshes, Maffittts, Scullins, Delaneys, Morrisons, Bakewells, Haleys, Thompsons, Taylors, Tomkins — when they are not Von Phuls, Von Shraders, Von Versens, Von Shrenks, and the like. *(None of these names, of course, being remotely Gallic.)*
- The new French: They do not need the tricolor boutonnière nor the tricolor hair ribbons to proclaim their birthright. Mrs. Max Gottschalk, Mrs. Giraldin-Farish, Mrs. Guy Studya, Mrs. A. Schreiber, and Mrs. G. Yorger all were indulging themselves in the free and uninterrupted use of the French language.

Scenes and Incidents at the French Fete de Charite Sketched by Marguerite Martyn

I approached them, and they did not even slow down to a pace considerate of any boarding-school "mastery" of the language. No, with one use of the phrase *"Comprenez-vous?"* they were off before I could stop them. One elderly man, seeing I was looking for a story, reeled off about a full column in length, and I just let him reel after managing to deduce that, stirring as it all must have been, it had happened to him way back in the French-German war of 1870.

The patrons, though wearing a gay facade, were serious over the purpose of this charity. More people were there than mere nationality would warrant, for it is said that only thirteen hundred native-born French live in all St. Louis.

Two women most intimately concerned in the present fate of France, Mme. Justine Robyn Gustine and Mme. Yvonne Le Glaive, captivated everybody's attention and deference.

Mme. Gustine's only brother is in the Belgian forces, and Mme. Le Glaive's young husband is in the grim trenches "somewhere in France," if he is alive.

In some minds, Mme. Le Glaive is a war heroine, even as her husband is a hero. At the first call from their native land, she urged him to join the colors. At that time she spoke scarcely a word of English. They had no savings, nothing. And she had not yet found a means of supporting herself.

"I may live, I may starve, but France needs you," she said. "We could never return to France, none of our friends would ever speak to us if you did not go now."

And he told her, "I do not like war, but my heart is French, and my country calls me." So he went over and fought until he was wounded, was invalided out, then enlisted again.

Mme. Le Glaive has employment as a seamstress in a large dressmaking shop. She awaits each European mail anxiously and hopefully, and these evenings at the *charité* event, she is a bright spot and a choice figure, attired as she is in a *pompadour polonais* and a Bréton cap, made of cotton print and bits of lace.

Conspicuous in the bazaar are some of the French allies. Mr. Seropyan, the Armenian, is one of the busiest merchants in all the little shops that border the dancing floor. Somebody accused him of introducing a false, even a hostile, note by having his cigarette vendors dressed in Turkish costumes *(Turkey, in the form of the Ottoman Empire, was an ally of Germany)*. But he said: "It is merely an Oriental costume, East Indian, if you will. We have our allies there."

One figure, so utterly a stranger and so distinct from the revelers, was leaning against a railing on a balcony, looking down on the gay scene. He wore a tightly buttoned reefer jacket and spring-bottom trousers and carried a sailor's cap. *You see him in the previous drawing, contemplating the festivities, his arms folded.*

He looked so wistful that I felt impelled to speak to him.

He said he was English (that was evident the moment he pronounced the word "milit'ry").

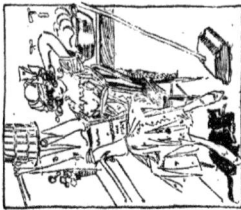

"I saw several years in the French navy," he told me, "and I've been seventeen in the American service. I have brothers fighting in France. It is a good cause, this."

• • •

On April 8, 1916, Martyn's fanciful layout (previous image) included a suffragette on a soapbox, a nosy neighbor and her quarry, a languorous woman catching up on the news from Mexico, a puzzled young lady seeking "powder," and others in a cartoonish display of how well prepared they all were to defend their country.

Across the bottom Martyn drew a "Preparedness Parade" which was puckish in its pretense "that women are already prepared for the horrors and responsibilities of war."

- *First in the bottom panel, that saber-bearing candidate for president of the city's prestigious Wednesday Club (for women) is ready to* "command any kind of battle."
- *Next, the woman boarding a streetcar on a rainy day and facing a phalanx of sharp-as-a-pointed-stick parasols must be* "prepared for the bayonet charge."
- *Third, the athletic girl who regularly climbs Art Hill just behind the museum is* "prepared to hunt" *revoutionary Pancho Villa, hidden away in the Mexico mountains.*
- *Fourth, Mrs. E.R. Kroeger, head of a committee working to clean up the city's noxious factory fumes and sooty household smoke, will combat poison gases with the funnel-shaped* "smoke consumer" *under her arm.*
- *Fifth, the model in the new bathing suit will wear it while dodging submarines.*
- *Next, a dressed-to-the-nines* "diner at a downtown café" *is* "prepared for the midnight surprise raid." *(Police had ordered that café music and dancing be stilled between midnight and 1 a.m. every morning. A pall of gloom fell over every restaurant and bistro in town which dared to stay open that late.)*

- *Finally, Mrs. E.T. Senseney, head of a civic campaign to get rid of flies, strides along with a fearsome fly-swatter that she could use "to swat Zeppelins" (the bouyant dirigibles of the German army).*

At top left of the panel (and next), uniformed men walk with their ladies past a feisty "reformer" who stands on a soapbox urging that women be allowed to do military service. "Why be colonels' ladies when we might be lady colonels?" she proclaims (a riff on Rudyard Kipling's poem, "The Ladies").

Next, a young woman (with dirty dishes in the sink) sits with legs crossed before an open window, reading a "Magazine for May." A cat lies beneath her chair, and a wheeled carpet sweeper (maybe a Bissell) leans waiting for use.

I ALWAYS SUSPECTED SHE WAS A SHIFTLESS THING!

THE PERISCOPE IS SURE TO PROVE A POPULAR PRACTICE WEAPON

BUT LOOK OUT THE WINDOW! She is being watched! Could it be . . . ? Yes, it is the top end of a periscope! (These devices were used to see above the embankments in trench warfare.)

One floor below, a busybody neighbor (in a spotlessly clean kitchen) mutters to herself, "I always suspected she was a shiftless thing!"

Let's return to Martyn's impish drawing of "Preparedness for Women":

In the third panel lounges a truly shiftless woman who reads that the principal needs of the troops in Mexico are chocolate (yummy), cold cream (sun protection), and cigarettes (cough). It occurs to her that she is well prepared with all three (she's got a whole box of bonbons on the table). She does wonder, though: "Horrors! I hope Preparedness does not mean conscription!" *(Not for another year, miss, not for another year. So gather up the men while they are still around.)*

Finally, upper right corner, a stylish young woman asks a puzzled shopgirl for some "Dupont" *powder, which she has heard about at a Preparedness meeting. It's* gunpowder, *miss . . .* gunpowder! *Not powder for the complexion.*

• • •

Next, Martyn heads for the border. Destination: Her husband.

7. In the Southwest

IN FEBRUARY 1916 MARTYN'S SKETCHES went missing from the P-D, so a reader wrote to ask: What happened to our favorite artist? The answer came briskly, in the letters column: "Marguerite Martyn is taking a vacation and will return shortly."

She went to Columbus, New Mexico, but this was before Pancho Villa's bloody raid that you read about in Chapter Five.

HE LEFT A TRUNK FULL OF WHISKEY, BUT THAT WONT PAY FOR THE MACHINE-YES, HE TOOK YOUR GUN NAW, HE WONT COME BACK— HE'S HALF-WAY TO FRISCO BY NOW —

YOU BETTER COME OVER AND GET RID OF THIS BOOZE THIS MORNING

CAFE

COLUMBUS, N.M.

IN ONE OF THE NUMEROUS NEW RESTAURANTS THAT SPRING UP OVER NIGHT.

Perhaps Clair and Marguerite spent Valentine's Day there. We know that they ate in a local café, because Martyn drew a crusty Columbus restaurant worker on the telephone (previous page), warning her listener that a customer had stolen an automobile and was headed for California, leaving behind a "trunk full of whiskey."

• • •

After that February break, it was back to work. Kenamore covered Mexican bandits and Martyn did the June 1916 Republican National Convention in Chicago.

In July, they were together again, in Texas. She sketched the El Paso plaza as an idyllic setting, with two soldiers chatting and a shawled woman waiting by the curb. A jitney bus could take anyone to the nearby copper smelter for twenty-five cents. (The three drawings were bundled in color on August 13, 1916.)

SMELTER 25¢

A CORNER OF THE PLAZA IN EL PASO

SKETCHED ON THE BORDER — JULY . 1916

The couple went into Juarez, Mexico, too, and saw this man and baby.

A TYPICAL SCENE

• • •

That time in Texas was a brief respite. Martyn returned home. She covered fashion shows, she sketched socialites. She interviewed politicians' wives.

In December 1916, Martyn and Kenamore were together full time, in Arizona. Neither of them had a byline in the Post-Dispatch for half a year, although Martyn did send back some of her work, in color, for a picture page on March 18, 1917.

Here's one of "Juanita of the Root Beer Stand" *in Nogales, Arizona, where border troops had been instructed to* "keep on friendly terms with the natives" *(following).*

JUANITA, OF THE ROOT BEER STAND.
THE SOLDIERS ARE INSTRUCTED
TO KEEP ON FRIENDLY TERMS
WITH THE NATIVES

Kenamore and Martyn were enjoying an extended vacation from the day-to-day grind, though I have a hunch that Clair was trying to write fiction and Marguerite was hanging around the University of Arizona campus in the company of other artists, and there were plenty of those in that sunny, colorful, brilliant land.

Their flat in Tucson at 1018 Lowell Avenue was two blocks south of the university, in an almost-empty field stippled with mesquite — exactly where the university's Dennis Deconcini Environment and

Natural Resources building is today. And there is no more mesquite in the beautifully tended campus.

In the winter holiday season, the two could have bundled up against the chill desert air and walked the few blocks north to the campus. There they could watch fourteen college women dance around a community Christmas tree, "dressed as dolls and representing every different character of doll known" (Arizona Daily Star, December 24, 1916).

• • •

In April the next year, the United States went to war, and everything changed. The vacation was over. The couple returned to St. Louis.

8. War Comes Home

BACK HOME, MARTYN AND KENAMORE both returned to print on Saturday, May 19, Martyn in the upper left corner of Page 3, and Kenamore in the upper right. I like to think that Managing Editor Bovard put them on the same page as a "Welcome home!" gesture.

RAMPANTLY PATRIOTIC

By Marguerite Martyn

May 19, 1917. Six months ago, St. Louis was so protestingly neutral that when I returned after a half year's absence, I scarcely recognized the old town. Patriotism and anti-Prussianism have exploded with all the more violence for having been restrained for too long.

I have been out in the Southwest, where war has existed for some time, where the military atmosphere permeates all walks of life. Every ranch house is fortified, and every citizen is a qualified marksman.

And they have female citizens where I have been. (*Arizona women had the right to vote since 1912.*)

Six months ago, we in St. Louis were laboring under the admonition of neutrality and discretion, and nobody dared to hint at favoring the Allied side lest some German sympathizer feel justified in protesting.

After a train trip of two days, I was awakened to the sight of a changed St. Louis. On countless flagpoles our national colors were floating to greet the morning. In the residential areas it was like passing through a lane of flags. At the station, every man, woman, and child tried to outdo the other in the novelty or flash of a patriotic buttonhole insignia.

There was as yet no draft. The government had to rely on volunteers. Women were called upon to rustle up men for enlistment in any military service. In St. Louis, they took to the streets aboard a procession of make-believe Navy warships, and Martyn joined in.

I am told I may do my bit with any branch — the Army, Navy, Red Cross, new Fifth Regiment, the Marines — they all need men. I will do a little with each, but first I chose the Navy.

The sailor is given credit for a sweetheart in every port, but where is the lass who doesn't love a sailor?

When I was in pinafores, I loved a sailor-boy doll in my collection, and there are jackies *(sailors)* aboard that make-believe "battleship" cruising our streets who look just like my beloved doll.

But, happily, they are much more alive.

We have two other "ships" in our recruiting flotilla — a "submarine" and a "submarine chaser."

When we weigh anchor twice a day from headquarters *(the Calumet Building at Seventh and Chestnut)* and set sail *(on wheels),* crowds gather in affectionate regard and curiosity. But we have accomplished little.

THERE IS ALWAYS THIS PERFECTLY USELESS BLOCKADE AROUND THE ENLISTING TABLE, BUT IT IS DOUBTFUL IF IT IS AN OBSTRUCTION TO RECRUITING

All of our blandishments are not bringing in very many recruits for the new American Navy. We average about 150 applicants each day, with about a fourth of them being accepted.

The results differ, depending on the neighborhood.

No Slackers in This Part of Town

Two evenings we dropped anchor at the corner of Twelfth and O'Fallon. This was Kerry Patch *(the center of Irish settlement),* and you might expect an audience here to be "scrappy," no matter the topic under discussion *(the Irish being stereotyped as combative).*

The original element that settled in Kerry Patch is now so diluted by "foreigners" of all nationalities that the Irish cannot be blamed for all the belligerency. On both nights, fistic disturbances showed that the old spirit has not changed.

The first: A man who failed to take off his hat during the National Anthem was set upon by every other man in the crowd. Only his swift pair of heels saved him from more varieties of punishment than I knew had been invented. (I was also initiated into the salty words of a sailor's vocabulary.)

The second: A rash onlooker was overheard using the word "slacker" *(one who avoids military service)* in relation to some others of the crowd.

"Nobody had better dare call me a slacker!" a listener roared in tones that put all the other orators out of business. "I've got my honorable discharge. I can show you the papers! I can show them to you, and it says on them 'excellent,' it does."

His wife fastened her arms around his neck beseeching him not to fight, but he shook her off, as well as several stronger arms.

Meanwhile, the first man was seen to weave his way toward the outskirts of the crowd, surely establishing a new record in fancy footwork. *(Next image.)*

"NOBODY BETTER DARE CALL ME A SLACKER

In our recruiting efforts, a band plays from the upper bow of our make-believe ship, and recruitment motion pictures are reeled off from the stern. A bluejacket gives an exhibition of wig-wagging from the turret. *(Using signal flags to communicate.)*

Young women "yeomen" and "petty officers" and other volunteers circulate through the crowd, distributing tracts and explaining the advantages of the Navy over every other branch of service.

I met with some unexpected types. An elderly woman with a shawl over her head was holding a group of children in some sort of spell.

"You have a son in the Army, haven't you, Mrs. Madsen?" said one little girl, as if by way of introducing the lady to me.

"Yes," Mrs. Madsen responded. She smiled as she looked at the motion pictures and tapped a foot in time to the music.

"Aren't you anxious about him now that we are at war?" I asked.

"No. He will be not be taken unless it is God's will, and, if so, it is the most honorable way in which he could go."

"But wasn't it difficult for you?"

"It would have been harder to have a son who hesitated, who might be suspected of fear. My grandfather fought for this country in the Mexican War *(1848)*, my father in the Civil War *(1860-65)*, my husband in the Spanish-American campaign *(1898)*, and my son has been with the regulars down on the border for several months. It is a satisfaction to me that the stock has not degenerated in this generation."

Proudly Proclaims He Is American

Perhaps there is not so much need to speak of preparedness within the so-called foreign sections of our city as we suppose!

A youngster not more than seven years old clambered upon the door of our conveyance and shouted in our faces, "I want to fight those Germans!" He spoke with a decided accent, and I had so little tact as to inquire, "What are you? Italian?"

"Dago?" he retorted, drawing himself up with dignity. "No. I'm United States!"

A total of twelve men signed up at the two meetings we had at 12th and O'Fallon. But it was a different story at Grand and Hartford streets, a lively South Side business corner, where we did not get a single signature.

It was a quiet, well-behaved, outwardly respectful crowd. Hats came off for the national air, and everything was peaceful.

But when we handed out circulars, they were often declined with the remark, "I have one." A man said, "Wait until we are really at war, and I might be interested."

My observations might seem pessimistic, but the results of these street meetings may yet fill up the Navy's ranks. We can't expect men to make a momentous decision in the hurried few minutes left of their lunch hour or in an evening when we unexpectedly descend upon them.

The parading-about is expensive, and officers are working under great strain and anxiety. But glance over the crowds and you see that many men are thinking it over.

It was inspiring when Major Wheeless of the Army Reserves finished speaking from the steps of the courthouse. One man in blue jeans fell into the ranks behind the squad and marched back to headquarters, where he signed up.

• • •

Kenamore's piece on the same page was a "thumb-sucker" — an article you could write from memory without having to dig up anything fresh. He introduced the recently appointed commander of the American army — and even advised on how to pronounce his name.

PERSHING AS HIS MEN KNOW HIM

By Clair Kenamore

May 19, 1917. His soldiers call Major-General John Joseph Pershing (accent on the first syllable) "Black Jack" when he is not near. The man is an erect, soldierly officer of about six feet, a strict disciplinarian, and one of the hardest workers in the Army.

Pershing got his nickname because as a young officer he had commanded the 10th Cavalry Regiment, the so-called "Buffalo

Soldiers," made up of African-Americans.

Planning is his specialty. He is not a man of attractive personality, and it is doubtful if anyone ever referred to him as "a bully, good fellow." He is a soldier in every instinct. The coldness of efficiency supplants any warmth in his makeup.

He does not talk much, and when he does, it is about the business at hand. He is a *(Missouri-born)* cavalry officer, and his choice to lead the American Army is a testimonial to the high regard he holds in the White House. He and President Wilson had probably not met even once before he was called to Washington last month.

• • •

Next: Hungry soldiers and unwelcome visitors.

9. Civilians Go to Camp

AT THE OUTBREAK OF WAR, the U.S. had about 127,000 officers and men in the regular army; there were an additional 80,000 federalized National Guard officers and men on the Mexican border, while 100,000 guardsmen remained under state control. To receive the new volunteers and draftees, the government expanded Jefferson Barracks, a dozen miles south of St. Louis.

This was a military site founded in 1826. Today it is a county park and is on the National Register of Historic Places.

Marguerite Martyn went to see how all those hungry men were being fed. Her report began in her sometimes roundabout way by mentioning a best-selling book.

MEN TAKE CHARGE OF THE COOKING

By Marguerite Martyn

June 3, 1917. H.G. Wells, in his popular novel "Mr. Britling Sees It Through," advances the theory that if women did the food preparation and cooking in the British Army, they would be more efficient than men.

I went down to the Jefferson Barracks to inspect an Army kitchen, and now I know that Mr. Wells's ideas are just fiction.

Captain McCammon forsook his desk and took me out for a tour of inspection.

First was the swimming pool, down near the *(Mississippi)* river bank, which will be finished by June at a cost of $2,500. It seems as spacious as any of the natatoria *(swimming pools, in a kind of pidgin Latin)* for which St. Louis is famous.

Then I "must see" the model dairy, he insisted. It is a concrete building, with lots of sunshine let in. It has modern milking machines, and the place is made sweet by plenty of running water. The forty cows were out in the meadow, but a disgruntled, pedigreed bull snorted us a resentful greeting.

"At the end of 'Lovers' Lane,'" said the captain, "is the piggery."

Avoiding the Pigsty

But the smell offended my sense of the romantic, and I offered to take his word for it that the post is rearing two hundred pigs.

Anyway, it was nearing noon, and we were getting an occasional appetizing whiff on the breeze from the other direction — from the mess hall.

It was in the kitchen that I saw how wrong Mr. Wells was about women superseding men as Army cooks.

Martyn kept scattering literary references.

I felt like a Lilliput among Brobdingnagians. *("Gulliver's Travels.")*

YOU FEEL LIKE A LILLIPUT IN A LAND OF BROBDINGNAGIANS IN THIS KITCHEN!

The tall, shiny coffee urns reflected such a Wonderland that Alice would have been content to stay on this side of the looking glass.

(You are correct. Lewis Carroll.) I can describe better with a drawing than with words how these giant vats made me feel. *(Martyn's back is toward us in her sketch, with an ever-present sketchbook tucked under her left arm.)*

Even with all the curiosity attributed to the female of the species, I could not have been able to see what was being cooked in those huge kettles if strong masculine arms had not lifted the heavy lids so I could peer inside.

I could not have begun to stir the hash with the paddles provided for that purpose, for they are about the same size as a rowboat's oars.

I could not have attempted to dish out the soup or the gravy, for the cooks use ladles holding a gallon on the ends of handles six feet long. The kettles made the jars used by Morgiana in the fairy tale shrink to a more normal volume, and there were enough of these huge vessels to serve up eight of the scalded forty thieves individually. *(More literature, Ali Baba and the Forty Thieves.)*

Some figures might convince you.

In the kettles were cooking 380 gallons of split-pea soup, and in the urns were boiling 500 gallons of coffee. Bread? 5,400 pounds being sliced. Meat? 2,800 pounds for two meals.

A large dishpan was overflowing with chopped onions, and sixteen 3-feet-square by 8-inch-deep pans were waiting to receive the hash once it had been mixed — in receptacles that looked like wooden rowboats, except the oarsmen stood outside and "paddled" inside. *(Sketch at right.)*

Gallant as any West Pointer, the man in charge of it all — Chef Stone — offered me a taste of macaroni or hash, and he was no less courtly when he acknowledged my praise.

He commands twenty cooks, the "officers" who worked among an innumerable regiment of privates, each in position and armed with trays of food, awaiting the order to go "over the top" into the mess hall to meet the assault of an army sixteen hundred men strong at this great recruiting post.

I wondered if the menial domestic tasks required of these soldiers were not discouraging to the ideals of bravery which inspire men to enlist. The captain suggested I ask one.

A man in a bespattered suit of overalls, standing idle for the moment, seemed a good subject, for he looked like an idealist, a thinker at least, with his broad, high forehead and shell-rimmed spectacles.

"Did you enlist with this end in view?" I asked, indicating the kitchen.

"Oh no! I enlisted for a commission," he replied. "I believe I can pass the intellectual examination. I have had two years in the University of Missouri, have served on the editorial staff of Kansas City and Denver newspapers and did social-service work for Montgomery Ward in Chicago."

A HIGHBROW
SUPERVISOR OF DISH-WASHING

Without a trace of apology, he continued:

"I am a supervisor of dishwashing here, and I am well satisfied, as are the several other college men here. We feel we are getting on the right side of Uncle Sam."

Soldiers seem to acquire a different attitude to the domestic service they scorned as "beneath" them when they were at home among their women folk.

I learned this when mess kits were passed out to a detachment about to entrain for further training "somewhere in the United States."

The interest and curiosity of the men was almost deferential when they examined the homely little outfit of frying pan, water bottle, tin cup, and cutlery.

Like Playing With Set of Dishes

They discussed and played with these utensils like so many girls with their first set of dishes. They read aloud the leaflet giving minute instructions for "such individual cookery as may be necessary for a soldier when thrown upon his own resources."

But we, their sisters, must not smile behind our hands when we remember how at home they used to look down on anything that was "girls' work."

"Was it your ambition to be a meat expert when you enlisted?" I asked a sergeant in the butchers' area. "Oh, no," he replied. "I signed up to be a soldier, but making the most of any detail is the business of a soldier."

I kept looking for some domestic service that women might render to replace men, but my feelings on that score were getting more modest. Potatoes were peeled by machine, and bread was sliced by one also. Dishes washed by machinery were done cleaner than if by any woman's hand.

At six o'clock, the meat shop had been scoured, sandpapered, bright and clean. It was the same in the kitchen, where the previous hive of activity was now all peace and quiet.

The Captain compared reports from his various officers and found that each man had been fed that day for just under 37 cents.

• • •

The Young Women's Christian Association erected a guardhouse to keep careful watch on the female visitors who wanted to chat up the men.

SIGHTSEERS ARE CHECKED

October 17, 1917. On a recent afternoon, I was one among the stream of visitors who stopped at the inviting little tent which, with its books and flowers, its chintz couch covers, and so on, had been made as homelike as women can do under the circumstances.

Miss Fox *(no first name given)* told me something of her mission.

"The instances of the young girl visitor who needs protection and those of the girl from whom the boys need protection are not always easy to separate at a glance," she said. "But a little conversation will reveal waywardness or wantonness if it is there.

"I tell the innocent ones that there is no reason a self-respecting girl should not see every point of interest about the post without an unpleasant experience.

"It all depends upon the manner in which she bears herself. She may be spoken to, probably will be, but then if she pays no attention, offers no encouragement, she is perfectly safe from molestation.

"To the other kind — when I have reason to believe that they are the other — I make a personal appeal. I remind them that these boys have offered their lives to keep the country safe for them. That they must help to keep them pure and to strengthen their ideals.

"Many times I have seen them depart without replying, with an expression of — to drop into the vernacular — 'Nothin' doing' written on their faces.

"And not only to the women has this tent proved a refuge," Miss Fox continued. "One evening just at dusk a youth came up from the woods and looked in wistfully. 'This is the time I usually spend with my mother,' he said. I welcomed him in, and after that he was here almost every night. Such a dear, lovable boy.

"After his departure, from somewhere in Texas I had a beautiful letter of appreciation from his mother."

More Help Needed

Miss Fox acknowledges that she cannot always be everywhere on the base.

"I cannot meet every streetcar and be 'at home' in this tent at the same time," she said.

"Many girls go by who ought to be questioned, and others need direction and chaperonage. We could use a large staff of hostesses. No doubt we will recruit others when we get a warmer building."

The shadows were getting long now, and a chill was creeping from the ravine below. Miss Fox busied about preparing her tea over an alcohol burner.

I was ready to depart, but she had to stay until 9 o'clock (when the post picture show and other entertainments close) and make sure that the sounds of "Taps" had dispersed the women visitors.

The next day Martyn sketched out a quartet of Women Who Were Not Wanted Around Here (one with a cigarette holder almost as long as her arm) and One Who Was Quite Welcome (a girl who was bringing a box of cookies baked by Mom to her older brother).

"AW, GEE, GIRLS, NOTHIN' DOIN'"

Y.W.C.A.

"THE GIRLS FROM WHOM OUR SOLDIERS NEED TO BE PROTECTED, AND —

THOSE WHO NEED A LITTLE DIRECTION OR CHAPERONAGE ARE OUR MOST FREQUENT TYPES OF WOMEN VISITORS."

The caption explains: "The girls from whom our soldiers need to be protected, and — those who need a little direction are our most frequent types of visitors." *The woman in the top panel has just been rejected by the YWCA guardian.* "Ah, gee, girls, nothin' doin'!" *the banished strumpet exclaims.*

• • •

As for Clair Kenamore, he went to Kansas, 375 miles away, to watch the Army turn civilians into officers.

10. Creating Officers at Fort Riley

THE GOVERNMENT HAD TO TRAIN an officer corps to handle all the new draftees and volunteers.

HOW THOUSANDS HAVE BEEN TRANSFORMED

By Clair Kenamore

July 29, 1917. In Fort Riley, Kansas, the nation is about to witness the outcome of an experiment which it took up three months ago, more in desperation than in hope.

When we entered the war, the government suddenly had to mass our enormous resources to make us able to help win it.

Specialists were trying to figure out questions of finance, munitions, transportation, and all the other demands of battle.

For manpower, Congress authorized a selective draft.

But how to provide officers for all those soldiers? The Army decided to train forty thousand men. They would have some military experience, if possible, but anyway would be men of education, character, and standing.

The Army chose sixteen posts to handle twenty-five hundred students each. It placed a trusted colonel or general in command of each, usually with twenty officers as instructors.

The men lined up in rows are being trained as officers.

It laid out a three-month course of study. It could do the job only with hard work, and maybe prayer.

The experiment is a success. Our nation's victory may hang on the speed with which we have prepared the officers who will lead in the field.

Of those who began training in Fort Riley, some two thousand remain. Four hundred will fail. Sixty-five can become second lieutenants, permanently, if they desire.

Each company will provide sixteen officers for the Quartermaster Corps. It was the original idea to make only officers for the infantry, but the demand of the vast and important work of the Quartermaster brought the change.

The Artillery wants forty men, and the Intelligence Department seeks ten.

Seeking Officers for Aviation

Seventy-five were transferred from Fort Riley after they qualified under rigid physical examination and agreed to enlist in the aviation branch of the Signal Corps.

Camp commander Colonel Tyree R. Rivers said:

"A test will come when the new officers take hold of their companies and make soldiers of the recruits. But the supreme test will be when they go into the field to lead their men."

The unqualified are weeded out. The man with a suspicious appendix is of no use. Nor is he with a bad lung, a leaky heart, defective and uncorrectable eyes or teeth that cannot pulverize Army hardtack. About three hundred of them have been sent home. They did their best, but they must serve their country in some other way.

The new officer knows now how to pitch a camp and where not to. He understands the rules for conserving the health of his command. He has a thorough knowledge of sanitation and hygiene and how to teach it to his men. *(Following photo shows a soldier looking dubiously at something new in warfare.)*

BARUCH
PORTABLE
LATRINE

As for battle itself, the bayonet fixed on the rifle is part of the daily workout, in temperatures that approach 105 degrees in the shade. The men charge out of trenches they have dug. They throw bombs into them, and they repel bomb and gas attacks. When the order is given, they go "over the top" with the same vim and dash they are expected to show in France.

Except in France the vim and the dash was to prove no match for a machine-gun bullet coming from the opposite direction.

• • •

Kenamore interviewed some would-be lieutenants.

CIVILIANS ARE GETTING TOUGHER

June 17, 1917. "The first two weeks were plain, unadorned hell," said the 31-year-old St. Louis office worker. "I had always been something of a beer hound, and combing my hair was the hardest work I had done for years.

"For a few days, I was sore everywhere. I had violent pains in muscles I did not even know I owned. But now, my stomach has almost disappeared. I hope forever.

"See that chest bulge out? Hit it once. Aw, hit it hard. Never fazed me. Feel that arm. Feel that leg. Just like granite, eh?

"Yes, the sitting-up exercises laid me flat for a while, but now — say, I'd rather miss breakfast than miss them. Well, I wouldn't exactly say that either; breakfast has become one of my favorite meals."

The next was a convivial, genial gentleman, addicted to work of no kind, who rated himself as an outdoor man.

"First," the soldier said, "it was the terrible dryness. I kept thinking about a scotch highball. That was the thing that made me so eager to get to France!

"Before the first week had passed, I was overcome by a great weakness and weariness. I was wasting away.

"One night I went downtown and weighed myself. I had gained five pounds. It was inexplicable. I took stock of my grievances and, by George, do you know that I did not have any! My aches and pains had disappeared. I was healthy and hungry and decided I had been a grouch and a pest, so I reformed.

"You saw me swing that old rifle in the Butts manual drill, didn't you? *(Manual of Physical Drill, United States Army, by Major Edmund L. Butts.)* No? That's too bad. You missed something mighty good. Another exercise you ought to see is me eating dinner tonight. You should write a piece about that. The sergeant in charge of the mess expects me to come out of this war as a fat brigadier-general."

Next chapter: How Americans pledged to feed this officer, and all the other servicemen.

11. Food Will Win the War

THERE WAS NO FOOD RATIONING during the Great War, but instead there was propaganda in favor of "conservation." So Americans had "meatless Tuesdays" and "wheatless Wednesdays." That took care of two days. But everybody was told to be abstemious on all the other days, too.

President Woodrow Wilson wrote to the country: "America and her allies must not run out of wheat, meat or fats. If we let that happen, Germany will win the war."

He appointed Herbert Hoover to head a national Food Administration to guard against that possibility. Hoover sent officials all over the country to spread the message. They asked "anybody actually handling food in the home" to sign a pledge card. It read:

"I am glad to join you *(Hoover)* in the service of food conservation for our Nation, and I hereby accept membership in the United States Food Administration, pledging myself to carry out the directions and advice of the Food Administration in my home, in so far as my circumstances permit."

Martyn wanted to see how this was working out, so she went —

DOOR TO DOOR WITH PLEDGE CARDS

By Marguerite Martyn

July 15, 1917. I was amused and enlightened, but I saw some pathos when I accompanied the women who are taking the Hoover food pledge from door to door in the congested districts. *That was the nice way of saying "slums."*

It's been thought necessary to make a complete canvass of the Fifth Ward, where many do not speak English or for other reasons would not be likely to understand or respond to the plea for voluntary signing of the pledge.

Volunteers went out from the Women's Committee of the National Council of Defense, and the St. Louis police furnished patrolmen as escorts and guides.

I went with Mrs. E.J. Smith of 1152 Walton Avenue and her escort, Patrolman Dennis Fogarty, when they did the south side of Morgan Street below Twelfth. They interviewed the women at the head of Italian, Jewish, and some negro households.

The news of our arrival was heralded from building to building and from floor to floor quite as if the telephone were not an unknown luxury and was in common use in that district. *(Next drawing shows three women using the balcony-to-balcony message app.)*

We found the government's ideas about food conservation much better understood than might be expected of women who sometimes needed an interpreter.

We found, without exception, a willingness to cooperate.

It was surprising how little discourtesy we met. There are some more moneyed neighborhoods where we would have been met with resentment — just opening doors and stepping into private rooms without being invited — asking questions that were decidedly personal, if not impertinent.

Mrs. Smith's speech went something like this, with modifications to suit special cases:

"We want to learn the number of people who eat in this house, so we can help Mr. Hoover in Washington regulate the food supply and prices. We want to know how much we will need at home so we won't make the mistake of sending more food to our allies than we can actually spare."

Waste food?! To say those two words together might have been considered a reflection on our sanity in such surroundings. Nor could we preach "thrift" to some of its supreme exemplars in this neighborhood.

Up several flights of rickety stairs laden with the accumulated dirt of ages we would find apartments which, though they were scant of furnishings, were as clean and sweet as scrubbing could make them.

In a rookery on Seventh Street fairly swarming with negro families, we found dark cubbyholes serving as kitchen, bedroom, living room — all in one.

But their darkness and dilapidation vanished with whitewashed walls, white coverlets on beds, stove and kettles shining, and attempts at decorations set out on bureaus.

These efforts to make the best of undesirable accommodations ought to cry shame to all the landlords of this district!

In a few cases, our offer of a pledge card — backed up with the vision of Authority in his blue coat, brass buttons, and badge —

brought an anguished dread to the face of a mother who assumed that our visit had something to do with the draft.

One old colored granny saw her daughter, the mother of a large family, sign the card, and she began to wail:

"Lawdy me! I knew they were going to take the boys and send them to France, but I didn't know they wanted the girls, too."

In another apartment lived a negro family with an unbelievable number of children and dogs. I spied a slice of bread, decidedly soiled, lying on the floor. When I called attention to it, the mother picked it up in a hurry. She had evidently come in contact with the food conservationists somewhere, and she knew her cue.

"I'll wash it off and make a puddin' of it!" she exclaimed.

But such misunderstandings were momentary. Mrs. Smith's ingratiating personality and our Irish officer's fund of diplomacy had their positive effects.

On Her Way to Work

An Italian girl on her way to work in a factory opened her door just as we were about to knock. She had read about the registration of women, oh, yes! *(Chapter 15.)*

"But I don't want to work in a foundry," she said. "I like the job I have now."

We found some cases which Officer Fogarty said he would report to the Provident Association *(a local charity that cared for the impoverished or homeless)*, and Mrs. Smith was

"I DON'T WANT TO WORK IN A FOUNDRY—

able to inform several hard-pressed women how they might earn some cash or food by working at the conservation canning factory on Fourth Street at St. Charles *(where volunteers, mostly women, put surplus produce from country farmers "into preserved form for consumption by soldiers.")*

Only One Meal a Day

One woman looked at the printed question on her card: "Will you take part in authorized neighborhood movements for food conservation?"

She said:

"It won't be hard for me to keep within bounds. We have only one meal a day now, after bread and coffee in the morning."

Officer Fogarty's method of getting answers to the most indelicate of our questions was to invite the householders to share what he made out to be a big joke.

"I notice you've all been too extravagant around here," he would say with a twinkle in his eye. "You're too lavish with your food. There are just too many beefsteaks and chickens winding up in your garbage pails."

And he would get the same answer, with some variations:

"We don't have any garbage pail. We have no waste. It's all we can do to get enough as it is!"

Mrs. Smith accepted that reply as equivalent to signing the Hoover pledge to conserve food.

Of course we had language problems, until some youngster volunteered to act as interpreter. And how precocious were some of these children! We met several who spoke Yiddish, Italian, and English.

And when we asked their nationality, they would say, "American!"

Altogether, not one person refused to take the pledge, administered in Mrs. Smith's sugar-coated doses, which made them privates in "Hoover's Army."

By October 1917, half of the nation's 24 million families had signed the pledge.

• • •

In St. Louis, the Women's Central Committee on Food Conservation (Mrs. Robert J. Terry, chairman) opened a "Hoover Store and Restaurant" at 410 North Seventh Street. Daily meals were prepared by two dietitians using substitutes for wheat, fats, butter, pork, beef, and mutton. A luncheon of cream of potato soup, corn dodgers (dump-lings), Boston brown bread, and fruit and nut salad, went for 35 cents.

This endeavor did not put other restaurants out of business, nor completely stifle the urge for wheat, fats, butter, pork, beef, and mutton, as you will learn in the next chapter.

12. The Autocrat of the Luncheon-Table

THE GOVERNMENT ISSUED A DECREE that restaurant coffee must be sweetened by waiters or waitresses and not by customers, who might be tempted to swipe the scarce commodity and take it home. And only one lump or teaspoonful of sugar! It was left to the local restaurant association to decide whether the customer could sprinkle the sugar on her fruit, or if the server had to portion it out.

'ONE TEASPOONFUL TO EACH CUP!'

By Marguerite Martyn

September 3, 1918. It may be that one of you diners just has to have two spoonfuls of sugar in your coffee, and it may be that the other really hates sugar in anything. So don't you think you could make a swap? That is, without any disrespect to the sugar regulations?

But, no. Here comes Hilda with the sugar bowl and the sugar spoon. *(In the drawing, the server has just unhooked the sugar bowl cover and stands ready to dish out the sweetener.)*

"One spoonful to each cup!" Thus she has read the order, and thus does she execute it. Precisely one spoonful goes into each cup, whether you want it or not.

Would either of you venture a protest with Hilda's dutiful, watchful, and suspicious eye upon you? Of course not.

You meekly consume your too-sweet or too-bitter coffee, consoling yourself with the thought that you and your table-mate are not the only war sufferers on this globe.

• • •

One of author Oliver Wendell Holmes's most popular books was "The Autocrat of the Breakfast-Table," from a series of newspaper articles narrated by a fictional New England snob.

Marguerite Martyn thought she would try the same tack, using as her avatar an opinionated St. Louis snob chatting with friends Alice, Carrie, Mable, and Blanche in one or more fancy restaurants.

In the first episode, the Autocrat (at left in the next drawing) alludes peevishly to the push to conserve food.

AUTOCRAT OF THE LUNCHEON-TABLE

December 4, 1917. "Yes, Blanche, dear, I AM a conservationist, and I ought to be lunching at the Hoover restaurant, but you need not look at me so reproachfully. I have been down at conservation head-quarters all morning working for others until now, and I owe myself some real food. *(Next image. Those dots on Carrie's face, toward us, are not blemishes but designs on her very chic veil.)*

" YES, OF COURSE, MY DEAR. , I **AM** A CONSERVATIONIST , BUT I'VE BEEN WORKING ALL MORNING FOR OTHERS, PREPARING BALANCED MENUS AND TELLING THEM WHERE TO SAVE. NOW I THINK I DESERVE A REAL MEAL."

"What do we *do* there? Why we show people how to prepare balanced meals and how to save. You may spare me your sarcasm, Alice — of course a balanced menu is real food. Although I admit that 'calories' do not taste as good as the word sounds. (*The headquarters were on the second floor of the Boatmen's Bank building at Broadway and Olive.)*

"Let's see. What's good here? Have you tried their lettuce sandwiches?"

The Autocrat hasn't got used to being a patriot yet.

"Only one lump? Girls, see what they offer us! Yes, of course it's all right; I approve of conserving. Hm. Do you suppose doing without it in coffee will help one become thin and sylph-like?

"Waiter, you will just have to bring me another lump of sugar. I'll pay for it if necessary. (*A fat tip will get around any wartime regulation.)*

How Many Girls?

"What was that you were telling me, Blanche, about those immoral conditions near that camp wherever-it-was? Five thousand girls, didn't you say? Or was it five hundred? Of course I don't believe a word of it, especially as it's been officially denied. Still, where there is so much smoke there must be some fire. *(Knowing giggles from all sides.)*

"Yes, I want dessert. How is this marshmallow nut paste with butterscotch sauce? Or this apple dumpling with hard sauce? Hm, they certainly don't Hooverize on the desserts at this place, do they? — although they don't bring the sugar bowl and the after-dinner mints any more."

The conversation shifts. The topic now concerns a garment not normally worn in plain view, offered to a girl whose fiancé was sent overseas.

"And because they were embroidered in the Philippines and he is an Army officer, out of patriotism her mother let her accept them? I never heard of anything so shocking. Why, when I was a girl, my mother wouldn't let me accept anything to wear from a man. Not even something to wear on the outside.

"And when I would break off an engagement, she always made me send back everything.

"Times have changed. Why, I would let my Marian accept anything, from an automobile to a house, and I would never make her send back a thing. Except the ring: that's the only gift that has any real significance."

She blots her lips.

"My, that was good. I could almost eat another. But one must have mercy on one's figure.

"Girls, have you tried the new yellow taxis? I took a ride in one the other day just to see how much they charge, though I wasn't going anywhere. Well, I thought I might need one sometime, and I didn't want to be a greenhorn. *(Following: Yellow Cab advertisement from the P-D of September 7, 1917.)*

• • •

"The checks they bring at the ends of these simple little luncheons are enough to give one indigestion, isn't it so? It does no good to utter a word of protest. They will just tell you it is the war and make you feel as if you were pro-German or something. *(A smirk.)*

• • •

"I asked the man who was fitting my shoes last week why the prices were so ridiculously high, and he said very loftily, 'Because our allies need the leather.'

"Meanwhile, he kept right on insisting I pay as much for cloth tops as for all-leather shoes. I got so provoked that I screamed at him, 'I will NOT pay twelve dollars for cloth-top shoes, so there!' And he said, still more witheringly, 'I can't do anything for you then, madam' and turned to wait on another customer.

"Bless your life, he didn't even put my old shoe back on, as they always do.

"So I had to sneak across the street and pay fifteen dollars without a murmur for a pair just like those he had asked twelve dollars for. What else CAN one do?

"Now, Carrie, it's *my* party. Here, waiter, give me that check! Oh, Alice, you've torn it. All right — Dutch treat today, but *next* time —!"

• • •

A few days later, the Autocrat is lunching with Mabel. They examine a menu.

MORE CHATTER AT THE TABLE

December 12, 1917. "What's the menu today? Wheatless? Meatless? Sweetless? Funny how those words all rhyme. We're getting close to eatless, it seems.

"Mabel, are you going to use your lump of sugar? May I have it? Thanks! I think I'll invite *you* for lunch every day.

• • •

"I see the suffragists on their way to Washington have made that pretty little Lucile Lowenstein spokesman for the Missouri crowd. I'm so glad. The prevailing idea that to be a suffragist you have to be old and ugly has always made me tired.

"I must say, though, that I am sometimes ashamed of the way women look and dress at suffrage meetings. I wonder if whoever appointed Mrs. Lowenstein hadn't had that same feeling."

• • •

The talk turns to an acquaintance and his wife.

"He says it will be just his luck to waste two or three months in a training camp and then have peace declared. But she insists on his doing it. And she is arranging to go over in the meantime as a nurse.

"She wants to outwit the government. Soldiers' relatives are not allowed passports, you know, but if she goes *before* he enlists, she'll be there already. *(A beat.)* She always had a mania for smuggling, too."

• • •

Because of the war, women were taking jobs previously done by men.

"I see that we are going to have women mail carriers. Well, I don't want any woman handling my mail. Oh, I just don't trust them, that's all.

• • •

"Have you a nickel, Mabel? Lend it to me. Thanks. I think a dime is enough to give the waitress while we are conserving, and it's so near Christmas time. Don't you? Besides, girls don't expect as much as the men waiters do."

• • •

Next the Autocrat is with Blanche, who asks about the Autocrat's morning.

OVER THE TOP

December 19, 1917. "My dear, I came downtown in the streetcar, and I was almost sickened by the smoke. One really needs a gas mask in St. Louis these days. I might as well have walked down, with all the energy I used stamping my feet to keep warm. That weird heating system they have on the cars gives the lie to the old saying that 'Where there is smoke, there must be heat.' It simply is not true.

"I've been trying to get some things to finish my kit bags (*supplies sent to soldiers at the front, both for morale boosting and for real need.*) One woman pawed all over the woolen gloves and socks and sheepskin foot warmers and said she could do better at the ten-cent store. It would be different, she said, if she *knew* the soldier she was sending them to, but she wasn't going to pay ten cents for a toothbrush and buy real woolen socks when she could buy imitations that looked just as good!"

The Autocrat must have had to pause for a sip of tea after such an outburst.

"Another girl convulsed us when she asked if the store had any sachet powder she could put in her kit bag, which was decorated with apple blossoms and blue birds! Wouldn't some Army man love her for that?

• • •

"No, we won't need to order bread. That's one of the economies of chicken à la king — unless they've begun to eliminate the toast."

All kinds of goodies were being "eliminated" on behalf of the war effort. The Wabash Railroad eliminated its morning passenger run between St. Louis and Chicago, "to release men and equipment for war transportation needs." Large electrical signs at night were eliminated to save on power. A German flag was eliminated from an ethnic display at St. Louis's Rose Fanning School.

• • •

"There was the drollest motorman on the car. He was chatting with a man who was leaning on the stove, 'to keep the stove from freezing up,' the man said! Don't some of those streetcar cut-ups just slay you?

"Anyway, the man was saying that when we get women conductors, the air would be perfumey instead of smokey. And the motorman said that a lady he knew could run the front end as well as the back, and 'us men can just sit around and let the womenfolks do the work.'

"It struck me that many a true word is said in jest. If women keep on assuming men's jobs in this town — one day postwomen, the next conductors and then traffic officers — where might it not end?

"Mind you, Blanche, I've always been a believer in equal rights for women, but there can always be too much of a good thing!

"Now, my dear, suppose we split an apple dumpling for dessert? *(Next sketch.)* They are always big enough for two. I've several presents to buy at places where I haven't charge accounts, and it's going to take some financiering to make my cash hold out.

"COULDN'T WE SPLIT AN APPLE DUMPLING FOR DESSERT? I'VE SEVERAL PRESENTS TO BUY AT PLACES WHERE I HAVEN'T CHARGE ACCOUNTS AND IT'S GOING TO TAKE SOME FINANCIERING."

"Goodbye, my dear. I feel much as I imagine a soldier must when he is about to charge out of his trench and go over the top. I am going back to tackle the aisles of that department store again."

13. Football and Poetry at Camp Funston

CAMP FUNSTON, BUILT ON THE Fort Riley grounds in Kansas, was named in honor of that general who, back in Chapter Five, was on the hunt for Pancho Villa. Frederick Funston died of a heart attack in January 1917, and this was to honor him.

It was one of sixteen camps hastily put together to mobilize a modern American army.

Major General Leonard Wood was the commander. He was a medical doctor who had been a standout college football player. He organized the Rough Riders in the Spanish-American War and was military governor of Cuba afterwards.

On September 5, 1917, draftees began to arrive in Camp Funston. They came from Kansas, Missouri, Arizona, Colorado, Nebraska, New Mexico, and South Dakota.

• • •

Clair Kenamore went there, about 375 miles from St. Louis, looking for local interest.

DRAFTEES RUN THE GAMUT

By Clair Kenamore

September 23, 1917. Eight thousand men drawn from all the white races of the world have poured into the nation's crucible which is

Camp Funston. Their only common characteristic is that they are between the ages of twenty-one and thirty-one. *(In those times, ethnic
origins in the U.S. were often referred to as "races.")*

It was the end of a long struggle in Congress and in the newspapers
for and against conscription. These were the first fruits, chosen from
among their fellows to fit themselves for a war on behalf of civilization.

The most picturesque trainload came from New Mexico. There
were many Mexicans *(that is, Hispanics),* bright and hopeful, their
faces lit with pleasant anticipation.

On that train also was an American *(that is, an Anglo)* youth with
a poet's face and great dark eyes like a girl's. He carried with him a
violin in a worn leather case.

These recruits bore a variety of clothing. Some wore useless
castoffs. Some shivered in Palm Beach attire. Others were in their
best silk shirts and silk socks and other brave raiment; those came like
bridegrooms, fittingly clad.

A man from Shannon County in Missouri brought his dog along, at the end of a rope.

With a Western Kansas contingent was another musician, seemingly from a small town. He carried a mandolin. He wore a green hat and green clothes which had pearl buttons on them wherever they could be put — cuffs, lapels, pocket flaps. In his tie was a small gold vanity pin which looked as if a fluffy-haired 16-year-old girl in a white dress had placed it there the night before the lad started away for the wars.

• • •

FOOTBALL THE KEY TO VICTORY?

September 23, 1917. General Wood has determined that the troops which he will lead to war shall be a division of athletes. He will train every man in athletics and develop to the fullest their powers in all physical development.

So thoroughly has the command at Camp Funston become convinced of the prime importance of athletic training that instructors are planning many diversions. Drills will be broken for a game of leapfrog, and as far as possible all other recreation will be of an athletic nature.

Lieut. Withington *(no first name),* who is outlining the campaign, expects to field five football teams from each of the 250 companies that will be here. This should show what a sporting metropolis Camp Funston is to become.

There is to be one marked difference from college sports. At Funston the men will learn all the dirty fighting there is to know: The cunning ways to kill a man, the quick strokes with knee or elbow, any way that a German can be put out of the way.

The officers figure that we are at war with the Hun, who knows no rules but to win and to whom sportsmanship is unknown. It is not considered wise to waste decency on the German soldier.

• • •

All the camps here are under similar regulations. Considerable comment has been made on the order that all officers must carry canes or riding crops; enlisted men will be "encouraged" to carry swagger sticks.

• • •

GENERAL SEES FUTURE SOLDIERS

September 23, 1917. In a wide greenhouse, surrounded by screened-in verandas, Major General Wood gazes over the expanse of Ogden Flats. Before him lies the vast Camp Funston, with the nucleus of the Army division he is to lead into France. (*Photo following, from the Arkansas Democrat, June 30, 1917.*)

General Leonard Wood.

He visualizes this division when it has been properly trained:

When a party of his men on leave passes down a street of any French town next spring, any American seeing them can say: "Some of our fellows. Yes, Wood's men. See the jaunty air? The swagger sticks? The truculent swing of the walk? Yes, Wood's men."

Or if a party of officers passes by, the onlooker will not be in doubt for a moment.

"Officers of the 89th," he would say. "You can always tell them. Note the chin straps, the general air of neatness, and they move like athletes."

Football on the Rhine?

And when the day comes, doubtless along the Rhine somewhere, when the games are played for the baseball and football championships of the Army, those unfortunate enough to have missed the struggles will ask:

"Who won?"

The answer will be: "Teams from the 89th took both events. Mighty tough fellows. Just can't beat 'em."

And, more important — when the high command casts about for a brigade or two for a desperate adventure, a perilous attack, or the holding of the line in a hard-pressed sector, where the strain will be long and the fighting hard, the generals will look around and say:

"Here is a place where we must give the best we have. This task is for the strongest in nerve and muscle and endurance. We will send them Wood's men. Send these orders to the 89th Division."

Talk about foresight, in athletics if not in warfare. Wood's 89th Division did indeed win the football championship of the American occupation forces in 1919, though the game was played in Paris and not "along the Rhine somewhere."

Wood was not there to see it. His Republican political leanings had ticked off President Wilson, a Democrat, so he did not lead the Army in France. As we learned in Chapter 8, that job went to General Pershing.

• • •

CALL FOR POETRY IN ARMY CAMP

November 18, 1917. Fifteen libraries supply the wants of the men in Camp Funston. And one of every five books that the recruits demand are works of poetry.

This percentage is from two to five times as heavy as you find in cities or at universities, the librarians say.

Rudyard Kipling is the most popular in that line. Then comes James Whitcomb Riley and the poems of Robert Service. The soldiers also want James Greenleaf Whittier, the *Rubaiyat,* Tennyson, and Longfellow.

The experts in charge — Willis H. Kerr of the Kansas State Normal School and V. Clayton of the Kansas State Library — cannot point to a reason for this poetical bent.

Aside from fiction and poetry, the demand has been for encyclopedias, atlases, French language, military handbooks, entertainment and games (especially Hoyle), the war, newspapers, maps, mathematics, the dictionary, religious works, and general history.

Since the negro drafted men have arrived, there has been a steady call for the poems of Paul Laurence Dunbar. *(He was the black poet who wrote a dozen books of poetry, four books of short stories, four novels, lyrics for a musical, and a play.)*

Classes Are Offered

The Young Men's Christian Association is sponsoring classes explaining our place in the battle, to give the soldier a knowledge of what he is fighting for and why he must fight.

There are other courses to teach both the French language and better English.

Among the negro soldiers the French classes prosper, and four excellent teachers have been found among the men.

Their popularity, according to a negro officer, is because many of the soldiers expect to remain in France. They have heard that the prejudice against their color which they find in the United States is unknown in France, and consequently France appears to them to be a very good country to live in.

More than eighty thousand African-Americans were taken in the first national draft. A Negro officers' camp in Des Moines, Idaho, turned out 104 captains, 397 first lieutenants, and 125 second lieutenants.

Enlisted men were trained in segregated camps throughout the country. Twelve thousand black trainees were sent to Camp Funston. Then they were assigned to two all-black divisions and sent to Europe. The 92nd fought alongside other Americans in the Meuse-Argonne campaign, and the 93rd was under the French army.

To make room for the newly arrived black trainees in a segregated area of Camp Funston, white Kansans and Missourians were taken from the 89th Division and sent south almost four hundred miles to Camp Doniphan, adjacent to Fort Sill, Oklahoma. These men formed a new division — the 35th. You will catch up with them in Chapter 17.

14. Women Step Up, Part 1

EVENTUALLY, TEN MILLION MEN LEFT their jobs and farms to join the American armed forces. Women moved into the breach.

(Versions of these three articles were published in "Marguerite Martyn: America's Forgotten Journalist.")

TRADITIONS SMASHED IN WARTIME

By Marguerite Martyn

February 16, 1919. Men went to war. Women went to work in the industries at home. Thousands of traditions were smashed.

Something like this great upheaval was needed to explode the fallacy that had developed with the machine age that women lack mechanical aptitude.

Miss Estelle Inskeep of 3213 Olive Street was one who ran a man-sized lathe at the Wagner Electric Company. She cut the largest metal parts in steel or brass and attained an output of about 750 pieces daily with not more than two or three "scraps." *(Image following.)*

And to prove that her skill was no mere adventure, when all of the mechanics began to feel they were speeded up beyond endurance and there were no pay raises to keep pace with an increasing cost of living, she joined with the rest and went on strike.

Months of arbitration have not reconciled Wagner Electric and their workers, so Miss Inskeep is now employed elsewhere. *(The "impartial" government arbitrator later went to work for St. Louis big business.)*

She ran a regular man-size lathe at the Wagner Electric Co.

But she has been entrusted by the union with several offices. And she went to the War Labor Board in Washington to present the employees' grievances.

"It is a satisfaction to know I can handle a lathe as well as a man, and it was fascinating work," she said. "But I have no desire to continue. See these burns on my hands from molten steel splinters, and this scar on my eyelid? I don't care to accumulate any more of them.

"I can make enough to support myself in trades that are not so hard on my nerves and my looks. I don't like to wear overalls; I like to wear women's clothes, and I do not care to sacrifice my complexion.

"Certainly, I don't want to take the job of any returned soldier as long as there is plenty of work that women can do and men cannot.

"But" — thoughtfully — "I shall always remain in the union and work for its principles. Every working woman owes it to the others to organize for better conditions and higher wages."

SHE DRIVES TRUCK SO BROTHER CAN ENLIST

May 31, 1917. All loyal nieces of Uncle Sam can quit complaining that they were not born nephews, able to offer their whole selves in the present great emergency. Not if they profit by the example of Miss Claudia Ellis.

When I told Miss Ellis the purpose of my interview, she said I could join her at 5 the next morning. But of course I was late, so we waited until the next day when she returned for her second pickup at 7 a.m.

I was her companion on the seat of the motor truck over a day's route, robed in the enveloping white linen ulster of the trade *(a long, white overcoat with a belt in the back)*. I became quite accustomed to our surroundings of several hundred pound of hams, bacon, pork chops, and sausages.

She guided the car flawlessly through the streets in the early hours and drew up at the rear doors of markets all over town. I learned what pleasant, sociable folks the market men are behind the scenes.

From her I also learned all the pet names and cuts of meat — dry salt, B'liner, bolo, rax, jowl, and so forth.

I came to see there isn't a reason in the world why women shouldn't be driving delivery motors everywhere.

A packing plant in the early morning is a weird scene, with men almost all speaking in foreign tongues, hurrying about in wooden-soled boots, and pushing great hand trucks laden with carcasses, or standing at vats slicing and hacking raw meat.

All along the route, the populace favored us with hearty signs of recognition. The visit of the woman meat vendor is in the nature of a social event. They joke and chaff and "jolly" and seem to find the keenest enjoyment in it all.

While we wove our way through the narrow ghetto streets *(the Jewish district)* and climbed the North Side streets near the river, I learned something about the girl who can drive a two-ton, 48-horsepower truck.

"I have my bit to do," she said. "I am determined that no foreigner and no slacker shall have this wheel. I am careful to see that any helper is always a boy under age *(and not a slacker)*.

"There will always be enough work elsewhere for men who are not eligible for the service, and I hope my pioneering will lead other women to see that we have got to be ready to replace the men."

WOMEN WORK TO BUILD AIR FORCE

July 11, 1918. Many St. Louisans do not know about the aircraft factories in our midst, so scrupulously has the government's injunction of wartime secrecy been obeyed.

At just one plant which I visited, from 90 to 100 women are employed in the construction of the precious planes.

St. Louis Women Building Airplanes

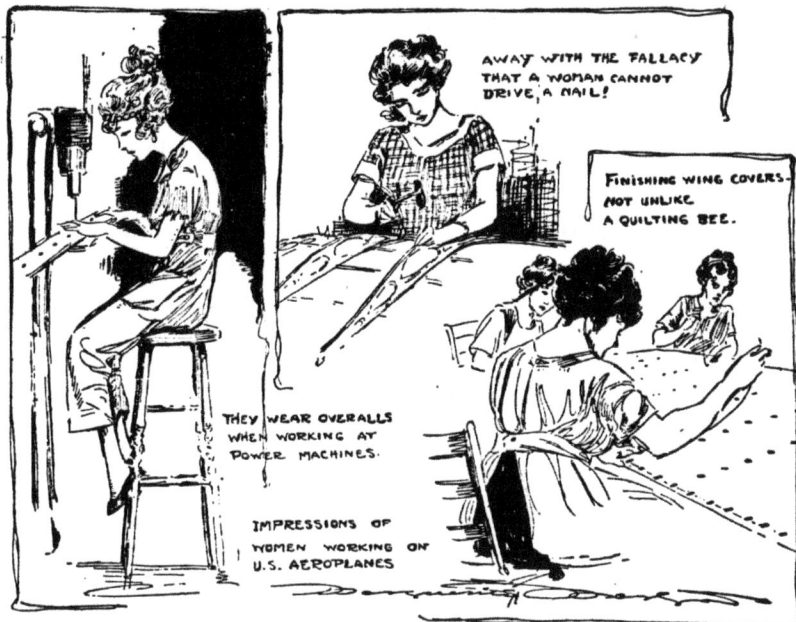

AWAY WITH THE FALLACY THAT A WOMAN CANNOT DRIVE A NAIL!

FINISHING WING COVERS- NOT UNLIKE A QUILTING BEE.

THEY WEAR OVERALLS WHEN WORKING AT POWER MACHINES.

IMPRESSIONS OF WOMEN WORKING ON U.S. AEROPLANES

President A.J. Siegel of the Huttig Sash and Door Co. was gratified at the spirit.

"They do not work as fast as the men, but they are most painstaking. For the most delicate processes, their fingers are better adapted than those of men.

"No, I don't hire them only because I can't get men, nor because I can get them cheaper; some of these women make as high as $27 a week, though they usually start at 19 or 20 cents an hour. I hire them because they can do the work and do it efficiently."

I saw the women in overalls or long aprons, stitching wing covers and "coping" them, binding struts with stout cord, varnishing landing gears and other wooden parts, greasing wires, riveting and boring at power machines, tacking, and glueing.

The strongest impression I brought away was the infinite care with which everything is watched. Piles of rejected parts lay everywhere.

A cross marked upon the defective piece indicates perhaps a tiny flaw that only a microscope might have revealed.

I saw an almost-completed airplane condemned because dust had blown onto its varnish before it was dry and had become ingrained. Why the finish should be considered when sometimes the life of a plane is so short, I could not see. But a workman declared, "Nothing is too good for our boys."

A large service flag floats over an inner court *(the blue stars indicating employees who had joined the military services).* The workers wear a pin awarded by Uncle Sam to show they are members of his industrial army. At the end of six months, they get to keep it forever, to pass on to their daughters.

15. Women Step Up, Part 2

IN MID-1917, THE FEDERAL GOVERNMENT took a survey of wom-en who wanted to register for war work. In another of Marguerite Martyn's frequent flights of fancy (the layout on the next page), she peered inside the befuddled minds of some would-be volunteers.

OPPORTUNITIES FOR WAR SERVICE

By Marguerite Martyn

July 26, 1917. The women's registration cards list more than 65 vol-unteer services. What a chance we have to exercise our dearest ambi-tions and talents!

The seated woman thinks, "I always have wanted to be an actress, but Daddy never would let me." *She dreams of herself in a slinky gown on a stage entertaining wounded servicemen.*

- *A corpulent lady.* "I can float fine in water. Why wouldn't I float in air?" *(She balances atop an airplane in her thought bubble.)*
- *A grandmother type accompanies her knitting with:* "I don't like overalls, but perhaps they'll let me wear an apron over them" *(as in her daydream of working in a factory, with an oil can).*
- *A smart-looking young woman with a long braid recalls,* "I did 'Maud Miller' in a play once and everybody said I was *such* a success, so I think I'll enlist as a farmhand." *(She sees herself wielding a hoe, but her braid is tucked in her hat, with ribbons as long as her hair. John Greenleaf Whittier's "Maud Miller,"*

about a star-crossed country lass, had indeed been made into a play. And a movie, in 1911.)

- *Next, a robust, middle-aged woman in sensible shoes muses,* "My children and daughters-in-law always told me I was a born nurse." *She dreams of herself as a young Florence Nightingale standing tall (and much slimmer), tending to wounded soldiers on the battlefield as bombs explode overhead.*

- *The woman in black sees herself stringing wire at the top of a long pole.* "My friends tell me I have a positive genius for telephoning, so I think I'll join the Signal Corps," *she says to her chum as she examines the list of job opportunities.*
- "Does it say anything about tank drivers?" *the friend asks.* "No? Then I guess I'll have to drive an ambulance."

STANDARDS FOR VOLUNTEER NURSES

April 23, 1914. How many husbands and sweethearts, brothers and sons have tossed aside their newspapers during the last few days and — with a girl-I-left-behind-me expression — looked down with masculine superiority upon a woman and announced that he was ready to follow the country's standard?

And how many women, goaded by her man's exclusive claim to patriotism, has retorted, "Then I'll volunteer as a nurse!"

Ah, my sisters, just being able-bodied doesn't make your services desired by Uncle Sam, as your menfolks' are.

Miss Jane A. Delano, Red Cross chairman of the National Committee of Nursing Services, is in town at the American Nursing Association conference, and she offers advice to the would-be war heroine. *(Photo from Burton Graphic, Kansas, May 21, 1914.)*

I asked her how a nurse can volunteer to serve. She replied:

"A Red Cross nurse must be a graduate of a school maintaining a definite standard and she must have two years' service in a general hospital, including the care of men, with an average of fifty patients a day."

She must have a license in those States where required, and she must be endorsed by her local Red Cross nursing section.

Miss Delano said:

"The United States relies on the Red Cross as a national reserve for the Army and Navy hospital departments.

"In case of war, all Red Cross nurses are subject to the national order. They are paid fifty dollars a month, which is half the wage she could get elsewhere, so the positions offer no financial attraction."

So, readers, you can see that once there may have been a demand (in novels, plays, and poems) for women in the melancholy service of nursing in wartime, but nowadays red tape and requirements have crowded out the romance.

It's been only since the San Francisco earthquake and fire in 1906, when confusion outstripped Red Cross services, that the organization has become more rigid and excluded temporary voluntary service, I learned.

Jane Delano died in France in 1919 and is buried in Arlington National Cemetery. Her statue there pays tribute to the 296 Red Cross nurses who lost their lives in the Great War.

16. Women Step Up, Part 3

IN TIMES GONE BY, UNIFORMED attendants operated a rheostat lever control within each elevator cabin in tall buildings — move it left for down and right for up. After the car settled into place, the operator would reach for a handle and pull the door open.

Marguerite Martyn was on the track of a story: A woman was actually running one of these things by herself!

LIFT OPERATOR HAS A 'WOODEN EAR'

By Marguerite Martyn

August 4, 1917. "Next car, please."

The order rang out from elevator Car Number One in clear, bell-like tones, gentle but firm.

Some would-be passengers waiting on the ground floor in the Third National Bank Building would have preferred that particular car, but a second glance at the operator was enough to persuade everyone to be herded into Car Number Two, as she demanded.

I ascended all the way to the top, the eighteenth floor, where I alighted and waited for Number One and its operator to arrive. Then I boarded, for the descent.

As the car dropped from floor to floor, she opened the gate, and the waiting passengers (mostly men) would hesitate before boarding. In some cases, a hat would come off and cigars would be lowered. In others, there would be an awkward indecision. *(The men hesitated because they were told by their moms to avoid smoking around women, even though they would like to smoke anywhere they damn well pleased.)*

MISS MABEL DIETZ.

A bold spirit broke the tension.

"Do you realize, Bill, that we are riding with the best-looking elevator conductor in the building?" And compressed masculine laughter exploded.

"In the building? You mean in the whole town!"

"Why, man, you'd be safe in saying 'In the whole country!'"

"Or anywhere else!" chorused the entire masculine cargo resoundingly through the echoing halls. There was a concentration of eyes upon a small, slim, blue-serge-uniformed back to see what effect such gallantry had upon the operator.

Not a very apt term, gallantry.

It had no effect whatever except that the owner of that uniform called out the floor numbers in an even more severe voice.

She is Miss Mabel Dietz, in charge of Car Number One.

First Woman Operator

During her relief period, we sat across a table, with two cool drinks. I was confronted by a group of exceptionally well arranged and regular features beneath a squarely set cap. There was a pair of fearless blue eyes, perfect teeth and the responsiveness of eighteen years of youth in a range of quick-changing expressions.

"If I paid attention to personal remarks and stares and compliments, I wouldn't be doing anything else all day long," Miss Dietz told me, "and my job would have vanished. So I just have to have a wooden ear."

She continued gravely:

"Yes, the safe conduct of the elevator is enough to take all my attention, for it isn't as mechanical as you might think, but I feel that the future of other women depends on me as the first woman operator in a public building.

"You see, Mr. Wright, my manager, is making a test case of me to prove that a woman can run elevators — a good many other building managers are watching the experiment. If I fail, it may mean that a good, suitable, and available occupation might remain closed to women.

"It is right hard sometimes to act like a wooden image in the face of smiles and compliments, and I didn't know there were so many nice, friendly people in the world. I haven't met a single grouch.

"The only trouble is that some are a little too friendly, if you know what I mean."

A Kind of Curiosity

She thought a minute.

"Of course I know it is because I am a sort of curiosity. After a while, when more women are engaged in the work, I won't be noticed, and then it won't be necessary to listen to personal remarks — any more than a woman working in any other place of public employment."

I asked about qualifications.

"Steady nerves and a level head. At any altitude. Some women get dizzy in elevators, and that would never do. It doesn't take much strength, but you must have good coordination and be able to think quickly.

"The most necessary thing at first is proper behavior, and I do hope that other girls who go into it will take the work seriously."

After a few nibbles at her ice cream, she confided further.

"Everybody is talking about doing her bit for the war effort, and I want to do mine, too. Then, another thing — I just love machinery. My father was a machinist, and he used to let me tag around with him."

Miss Dietz's mother is a widow.

"I was a telephone operator before this, and I got a lot of fun out of experimenting with the wires.

"And I just love that elevator car of mine. It behaves beautifully under my control; I have tried all the others, and none is as nice as Number One."

Faith Is Vindicated

Guy H. Wright, the manager of several buildings, insists he is a crank on the topic of elevator efficiency. He is the pioneer in substituting women for men as conductor.

"I have always believed that women would be ideal in this position because they will listen to instructions," he said. "My faith is being vindicated. Miss Dietz is saving me money in current, in wear and tear on my copper wiring and the carbon contacts by obeying the simple injunction to close the door before starting the car — something I've rarely been able to get a man to do at all times.

"I was afraid there might be some trouble about the draft blowing a skirt in contact with the current, but we've experimented with skirts of both light and medium material, and I find there is nothing to object to.

"If eventually we have vacancies created by the war, we know now that we can substitute women for men in this work."

• • •

Charlotte Rumbold (next photo, by Kaji-wara) had been St. Louis supervisor of recreation for many years, credited with bringing about vast improvements in the city's parks and playgrounds. Nobody doubted that. Yet in 1915 when she asked the city fathers for a pay raise to bring her up to the level with other administrators (all men), they refused.

She wasn't a voter, they said, so they wouldn't do it. (This was Catch-22 before the term was ever invented. Women did not have the right to vote in Missouri at that time.)

Rumbold immediately quit and soon got a better-paying job with the Cleveland Chamber of Commerce in Ohio. Three years later, she was back in St. Louis, attending a conference on city planning.

RUMBOLD SKETCHES LIFE OF WOMEN WORKERS

May 29, 1918. Friends familiar with Miss Charlotte Rumbold's years of effort to alleviate social poverty in St. Louis, who have always praised her work with our local playgrounds system, public baths, and other recreation, will of course recall the administration's failure to reward her with a man-sized salary.

They will be glad to know that many of the dreams she had for St. Louis are materializing elsewhere, and she is being rewarded accordingly. Her woman stenographer in Cleveland is said to be earning a higher salary than Miss Rumbold was paid here.

I ran into her quite by accident yesterday, and she gave me a lively account of the branch of war work she is engaged in at present.

Equality for women is her predominant ideal. She has been most interested in how women are affected by the new occupations they've entered since the beginning of the war.

She said:

"Working at certain jobs in which a man made three or four dollars a day *(paid at piece rate, so much money for each part finished or each job accomplished),* women now are making ten and twelve dollars at the same rate of pay.

"The reason is not that men were slackers but that women work faster. Their fingers are nimbler and not so clumsy in working with the small tools and machines employed in so much of the munitions-making. Quick, nervous women are excelling at these tasks.

"For the heavier work, strong-bodied, foreign-born women, accustomed to manual labor, are chosen, and often they, too, do better than men."

New Economy for Women

She gave an example.

"I was watching a woman operating an overhead crane in a steel plant, and I asked the overseer what the woman might do if something went wrong with the machinery.

"'What would you do in such a case?' he asked me in return.

"'Why, I'd probably scream and sound all the alarms,' I responded.

"The overseer replied: 'That's exactly what that woman would do, and what she should do, whereas a man would probably attempt to find where the screw was loose and mend it himself, in the meantime upsetting everything and maybe endangering life in the delay.'"

Miss Rumbold was enthusiastic about the new economy for women.

"At the Remington Arms Works I watched one stream of women pouring out of the shops at shift change while another was jamming the entrances.

"I singled out one worker and asked her some questions. She told me she had been a schoolteacher but was now making three times as much.

Joins Club, Takes Trips

"'Don't you find your associates sometimes uncongenial,' I inquired.

"'Most of the women I work with are very nice,' she replied, 'but I don't have to depend on them for society. I have bought a Ford, and I drive to New York on holidays *(about nine hundred miles).* I was able to join the Women's University Club, which I never could afford to do on my old salary and assume social obligations that I couldn't afford before.

"'My friends, especially my men friends, are more interested in me now because I have so many new experiences to talk about!'"

Miss Rumbold asked the worker: "How do you think men will feel when they come back and find their jobs taken?"

She was answered: "I'm afraid the men won't get all their jobs back. The women won't want to give them up, and the employers will not want to give up their new workers."

She Will Buy Shoes

What about the moral effect of these changes on the women? Won't it interfere with their destiny as wives and mothers?

"It will not change feminine nature," Miss Rumbold said. "The first thing a girl does with her twelve dollars a day is to buy a pair of high white shoes. Then she may complete her wardrobe. Next she will likely buy a Liberty Bond.

"No, I don't think the maternal instinct, which is stronger than almost any force on earth, will allow her to refuse to marry and have children. But she is going to be more choosy about whom to marry, and she will certainly have her own bank account afterward!"

Then she said:

"However, I have noticed an inclination among these girls to speak scornfully of the marriage bonds, though they by no means manifest a scorn for men. And a girl and a man, both with their pockets full of money and sudden unlimited freedom, are a dangerous combination."

"You mean it is likely to lead to a condition known as 'free love'?" I asked.

"It is going to lead to a great deal more freedom of choice among the women," she declared. "A strange part is that the girls do not seem to lose caste among themselves for their boldness as they once did.

"How conditions will adjust themselves," she mused, "I cannot predict. It may mean a revision of moral standards with men and women more nearly on an equal plane, or with women dictating standards where men have always dictated them before."

• • •

"She's a peach."

In the slang of the early twentieth century that means she was attractive, easy on the eyes. Marguerite Martyn drew her for the P-D's September 9, 1913, issue.

THE CANNIBAL, OR, PEACHES

Drawn for the Post-Dispatch by MARGUERITE MARTYN.

But five years later, there was a war on, and —

STENOGRAPHERS, NOT PEACHES, ARE SOUGHT

June 9, 1918. "Peaches" are not wanted as Red Cross stenographers in France.

This fact was made clear at an examination of women conducted by Mrs. Ruby Lester Fleming, who is on a nationwide tour to choose a staff of stenographers for the Red Cross offices in Paris.

She is receiving applications at the Red Cross headquarters in the Railway Express Building.

Most of the applicants I saw waiting to be interviewed were of that familiar type with very short skirts, very high heels, very sheer stockings, very thin blouses, very coquettish hats and, in almost all cases, jewelry bearing the insignia of some branch of the military.

These with the jewelry were the sweethearts, wives, or sisters of soldiers already in Europe. Regulations forbade the women from being sent over to follow them.

Not all were young, but it was not just the faded ones who indulged in the artifices of rouge and peroxide that are intended to produce the semblance of youth.

All who were examined while I was present were turned down, though I was told that some five or six applicants in St. Louis have been found eligible.

And Mrs. Fleming did give high praise to Misses Gertrude Chapman and Martha Finnell of our city, who are already at work in Paris.

Some Applicants Will Be Accepted

Three recent applicants have already progressed beyond the first stage. They are two young women who have conducted offices of their own — and a woman lawyer.

I encountered the lawyer in the waiting room, and she said, "I am willing to do anything, but I do hope they won't ask me to do filing."

Doubtless out of patriotism she will overcome this aversion and the Red Cross will gain one of the high types of women that it seeks.

One little girl with a lame ankle seemed to satisfy Mrs. Fleming. She was able to show that she had passed the test for service for the Navy until it came to the physical examination. Her patriotism was attested by her records as an officer in the Red Cross and Liberty Bond campaigns. But she will also have to pass a physical examination to be sent to Paris.

Mrs. Fleming told me that one glance at a candidate usually is sufficient. A suspicion of rouge is an immediate demerit, and there are other surface disqualifications.

Two girls came in with arms linked, wishing to be examined together. This being forbidden, both wanted to know if there would be any objection to two chums going together as far as Paris and then through the danger zones.

Another girl (with touched-up eyebrows and lips, a skirt about fifteen inches off the floor, and a hat that could never be packed in a steamer truck) said she had two brothers in the military. She knew about the rule against soldiers' relatives going over, but she thought that since the Red Cross was so hard up for stenographers, well, maybe the regulation could be overturned.

Crisp and Efficient

Mrs. Fleming's own personality should discourage ineligible women from applying.

She is crisp, businesslike, and efficient. Her complexion is clear, smooth olive. Her hair waves naturally back from her forehead. Even her eyeglasses pinching her nose do not destroy its exquisite modeling.

A well-tailored Red Cross uniform, with its insignia of staff officer and service stripe, worn over a khaki silk shirtwaist, with a snug collar and cravat, sets off to advantage a softly rounded figure. *(Photographer Takuma Kajiwara made her remove her glasses. She looks like a peach in his photo.)*

"We want mature women, preferably between thirty and forty years old," she said. "The first requirement is loyalty, patriotism and sincerity of purpose. In this work, there is never-ending strain, so we must be sure of our staff, mentally, physically and temperamentally.

MRS. RUBY LESTER FLEMING . . .

"We were under fire during the long-range bombardment of Paris. It is less spectacular to say that you stuck to your typewriter than you stuck to your ambulances filled with wounded, but I assure you it requires the same cool nerves.

"I am glad to say that not one of our staff faltered in her work, and that is the kind of woman I am looking for."

She added:

"Even a woman who is accustomed to high-class executive work must go with a willingness to do the most rudimentary tasks of clerkship if she is assigned to it."

A French visa is necessary on the passport, Mrs. Fleming said, and "France is suspicious of a German name; you cannot reason with a country whose heart has been so torn."

Live and Work in Paris

Successful candidates must remain for a year.

"Much of the work is supplying information to the families of soldiers on the casualty lists. Thousands of inquiries come in; letters have to be written that will relieve the anxiety of relatives."

What about living conditions?

"The workers can live as they choose. They are paid what they seem to require for their expenses; it costs Americans at least $120 a month to live in Paris."

Mrs. Fleming is a product of the wartime call of patriotism that has placed women in the jobs previously monopolized by men. She left a high-salaried executive position in New York.

In her efforts, she is not recruiting a makeshift regiment as much as she is weeding out and selecting the highest type of woman from the unlimited applicants available.

Meanwhile, Clair Kenamore had been following the training in the hinterland, Camp Doniphan, Oklahoma, where real regiments were being put together (next chapter).

17. Building Soldiers in Camp Doniphan

THE HONOR OF MISSOURI WAS severely tested when the Army decided to wipe out Company D of the State's National Guard and distribute its remnants among other units. Missouri Governor Fred Gardner made a complaint to President Wilson, who met with three Missouri congressman and promised to speak to Secretary of War Newton D. Baker. Nothing changed, of course.

The company had been called up for service on the Mexican border (Chapter 5). It didn't make much news there except when (1) a local judge gifted it with a hunting dog as a mascot, (2) its band learned a new regimental marching air, "You Gotta Quit Kickin' My Dog Around," and (3) a guard killed a drunken soldier.

Earlier, Company D had been noteworthy for (1) seeking to be sent, but never going, to the Spanish-American War, (2) winning awards and accolades for drilling and parading, and (3) dishonorably discharging the son of ex-Missouri Governor John S. Phelps after the young man painted a companion black from head to foot while the fellow was asleep.

This merger of the National Guard unit with the regular Army did not go smoothly. The rancor lasted through the fighting in Europe (Chapter 29).

MILITARY SHUFFLE IS PONDERED

By Clair Kenamore

October 7, 1917. So far as I can see, in the shakeup some officers got what they wanted and others lost what they had. The winners were exultant and the losers were depressed.

Two have resigned, but certainly the War Department does not look with favor on resignations at this time.

The Second Missouri was the oldest in the State, the largest, and frequently accounted the best. It has been cut up into machine-gun companies and the identity of the regiment in lost

As I come from South Missouri, I speak the language of these men, so I sought out three buck privates who sat in the sun and asked them what they thought.

"Well, it's pretty tough to lose the old officers," one said, "but we kept a good many so we'll get along. It was a cinch they'd pick us for machine gunners; we were by far the best regiment in the State and the best shots, and there's no question what we'll do. Watch our smoke: Next spring we'll be playing 'The Houn' Dog Tune' on those guns."

Uplift and Thrill Among Trainees

Airplanes whirr overhead, a great sausage balloon bobs on its cable, six hundred men are throwing a line of trenches across the slope of a hill, a thousand squads of eight each are drilling, motorcycles raise clouds of dust, and guns boom.

There is uplift and thrill. It is fine to see a mile of field artillery winding across the plain, and there is inspiration in the gallop to position, the whisking of guns into line, the quick withdrawal of horses to safety, a word of command, and the climax as the guns begin firing.

The sound of the airplanes is in the key of war. The thunder of the truck trains, the sputter of the motorcycles, the music of the bugles, and even the howling of the dogs are parts of the symphony of war.

There is inspiration, too, in the spectacle of the captains and their men working with pick and shovels to dig and equip trenches under the command of Lieutenant July Champenois of the French army.

Major General William Wright; Lieutenant Colonel R. McCleave, his chief of staff; and Captain P.C. Kalleck, his aide, are all in France, learning the latest tricks in their trade of war, and Brigadier General L.G. Berry, an artillery officer, is in charge.

Back in Camp Funston, General Wood had thought that football games would be a great way to train soldiers for war, but General Berry scoffed.

"These men will be ready for France when their sixteen weeks are done," General Berry said. "I fear we will not have much time for football. Every effort will be made to harden the men in the least possible time, and this can be done only by exercises in which every man takes part. Football is intensive athletics by a few as entertainment for the many."

So much for that idea.

There are not enough rifles here to arm everyone, but each company has a sufficient number to give everyone a chance at the manual of arms and the drill. *(The country was unprepared for a major war. Soldiers made mock rifles from slabs of wood.)*

Battles From the Past

Camp Doniphan is in a wide valley of the Wichita Mountains. Adjacent is Fort Sill, built when it was the business of the soldier to fight Indians. There is a new, modern fort also. Here are the hangars of the aviation school, the balloon tent and the wide landing field.

Dust devils dance over the prairies where the soil is dry and hard, the mountains are bare, the air is clear and the trees grow by clumps in little hollows. So favorable are the conditions that it was not until recently that the tents of the enlisted men got any flooring. One good rain would have spoiled it all.

Shrewd-eyed men are high above as artillery observers in airplanes. Some of the students here claim that those pilots have "retired from active service" to the security of a quiet job. *(Not true. Those rickety crates were constantly being shot down during the battles to come.)*

At Fort Sill, young men are trying out the gas chambers *(where trainees learned how to don gas masks)*. On the hillside, infantrymen learn the clumsy, underhand delivery by which the hand grenade is landed in the enemy's trench. *(The Army soon gave up that softball way of lobbing a grenade and began pitching it overhand, as every boy did on the baseball field.)*

Doctors do rigid physical examinations. Some specialize in diseases of the mind; their business is to see if any man is unsuited for the occupation of war.

The photo shows soldiers lined up for typhoid shots.

LINE UP FOR TYPHOID INOCULATION
CAMP DONIPHAN, OKLAHOMA

The training began September 1 and will last sixteen weeks, so by January 1 the men will be ready for the transports.

Did Camp Doniphan do its job? Maybe not.

A senator complained that conditions at the base hospital were "nothing short of deplorable"; *two officers were court-martialed after a soldier died of spinal meningitis. Commander Berry was called out in a scathing postwar Army report (October 26, 1918), which charged that his division* "was not well trained and fit for battle." *(You learn more about this in Chapter 29.)*

Nevertheless, as these men were being trained for service on the front lines, whether for good or ill, women back home were preparing bandages to bind their wounds.

18. Sewing and Kissing Will Help Win the War

BANDAGES FOR THE BATTLEFIELD

By Marguerite Martyn

November 3, 1917. Wanted: Fifteen hundred women! And they need you right away!

The call for women's service in this war was brought home when I visited the Red Cross surgical dressings school at Washington University.

This was a place where women could learn how to teach others about preparing and folding bandages.

Rush orders for more gauze packings and pads and special dressings are streaming in, and though they are being turned out at a rate of sixty thousand a month, that is still not enough.

I saw two women fairly running from one end of a twelve-foot table to the other, stretching lengths of a gauze to be cut later by machinery. Others dashed off with armloads of the stuff in response to hurried calls from heads thrust in at doorways.

In the top panel, next drawing, a woman cries, "Come on! More gauze — quick!" The other responds, "But, my dear, you've had about three tons of it this morning!"

Next, an attendant moves a cart filled with bandages while women behind her make packs "as dainty as Christmas packages."

At bottom left, a trainee has run into a problem with her electric-powered sewing machine. "It isn't sparking right, or it's missing on one cylinder, or something."

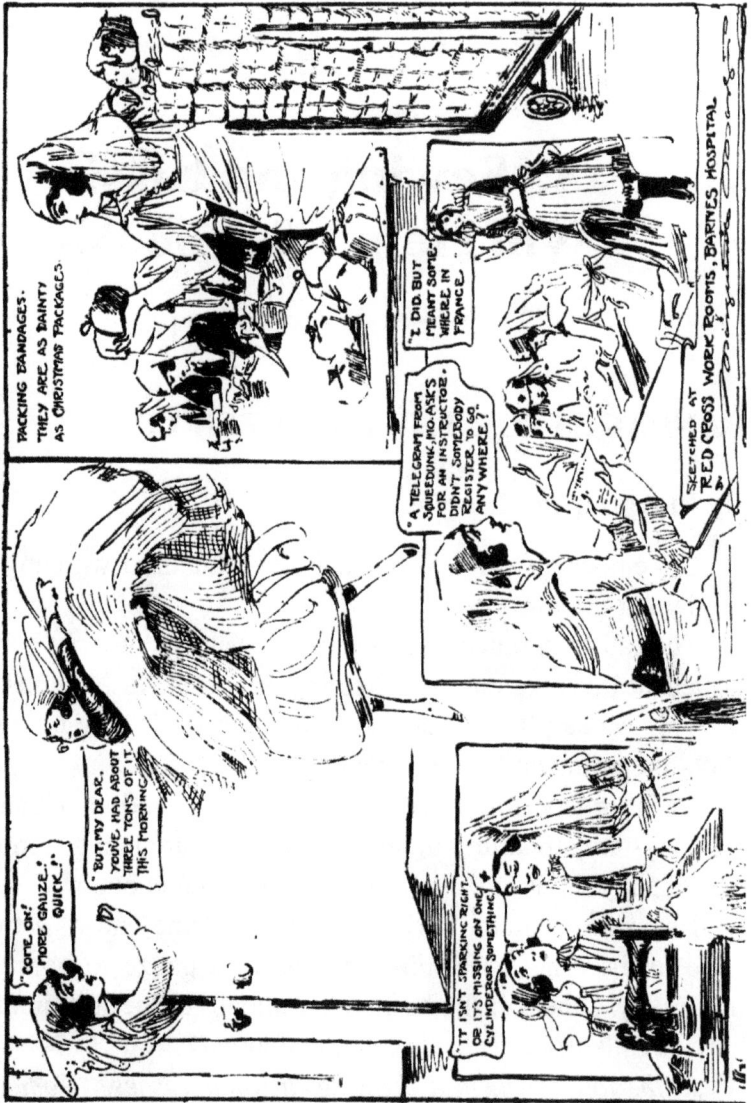

Finally, a supervisor reads a frantic telegram from Squeedunk, Missouri, (not a real place) asking for an instructor in bandage-making. "Didn't somebody register to go ANYWHERE?" *she calls out.* "I did," *comes the response,* "but I meant somewhere in FRANCE."

Vigorous Exercise

At the cutting table, Mrs. J.A. Haskel and Miss Johns spread seventeen bolts of gauze in a morning's work. Each takes an end of the fabric and runs to the opposite end of the table, hooking it thereon. Then back again, over and over until the bolt is exhausted.

And other women were hurrying around the same way at other tasks. *This is why automation was invented.*

These rooms can house fifteen hundred workers at a time *(that's a lot of volunteers!),* but when I visited on Wednesday morning, only half the chairs were occupied. Wednesday and Friday afternoons are even worse.

I recognized some of the women as more familiar behind the wheels of their motor cars than behind a sewing machine.

COMPOSITE VIEW OF THE MOTHER SPIRIT

November 7, 1917. When you pass the corner of Tenth and Locust streets and look through the wide expanse of the windows of the Red Cross sewing branch, what do you see? *(Next drawing.)*

I see a "mother face." The mother who no longer fibs about her age, who isn't ashamed to let the gray hairs show and who assertively wants you to know she is old enough to have a son in the Army.

Sometimes she is a proud mother with courage. Sometimes she is anxious. But always she is busy, her fingers flying over the coarse muslin or homely flannel garments she is fashioning.

I have the impression she has left her family's dinner steaming on the fireless cooker and the front-door key under the mat for the schoolchildren. She will rush home after work to see that the fires are burning and the lights glowing to welcome her homecoming menfolk.

She hasn't time for sad thoughts about the war, and she must not yield to melancholy — for if she does, all those who depend on her will weep also.

While she sits and sews for the soldiers, she permits herself only the rarest of worries about her own soldier boy. Or of the other mothers' sons who will wear the garments she is finishing.

Composite View of the Mother Spirit in War Time
Drawn for the Post-Dispatch by Marguerite Martyn

Oh, the tender, the brave, the ever-faithful "mother-thoughts" which are sewn into those seams. Aren't they enough to keep our boys from harm and to overthrow the evil we are struggling against?

When we look through those windows, we look at the mother-spirit of our land.

ITALIAN OFFICER 'REWARDS' ST. LOUISANS

July 20, 1918. A bit of romantic Italy was transplanted to the Italian Red Cross surgical dressings shop at 1230 Olive Street when Lieutenant Roberto di Violini made a visit.

The shop has been operating for two weeks and has turned out more than a thousand dressings. Because these supplies are going straight to the Italian front, many leading women among the twenty-five thousand St. Louisans with Italian names are supporting the work.

Di Violini awarded insignia to women who qualified by thirty hours of service as experienced workers.

He pinned a red, white, and green ribbon on the cap of each woman, then planted a kiss on the cheeks of each. He had prepared in advance by explaining the gesture was an Italian custom to accompany a decoration. *(Italy had been in the war against Germany since August 1916.)*

Five women were honored — Misses Olympia and Theo Monti, Mrs. Edward Monti, Mrs. Angelo Ballo and Mrs. Madison Whiteside.

They accepted with grave dignity, only a flush deepening the color of their cheeks, while onlookers of a less sentimental race gave themselves up to cheers and applause.

When Mrs. Whiteside approached for her decoration, the officer, quickly perceiving that she was not even remotely a countrywoman of his, delicately refrained from bestowing the salute.

Theo Monti, at left, looks suspicious as Lieutenant Di Violini prepares to plant a kiss on her mom, who seems to welcome the idea. At right, Mrs. A. Ballo, who has just been "decorated," is woozy.

The lieutenant, as an officer of Alpine troops, wore a smart gray uniform with a collar bearing a strip of green and a star. On his right sleeve, two stripes indicate wounds. His hat is a picturesque affair, with a rolling, upturned brim, through whose band he had thrust a feather.

He served on the Piave *(river)* and is now on a mission to encourage greater fraternity between us and our Italian allies.

He speaks English fluently and floridly, with occasional amusing lapses into American slang — which he learned from the men of his regiment, some two-thirds of whom had lived in the United States and some of whom spoke mostly "American" and little Italian.

At the beginning of his talk, he proposed three cheers for "those big American boys who are doing such good work on the battlefront today."

Everywhere he goes, he reads a pledge and asks the workers to subscribe to this compact of fellowship to banish tyranny and autocracy from the Earth.

Di Violini continued his tour through the East and Midwest, bestowing kiss after kiss wherever he went.

19. Knitting Will Help Win the War, Too

EVEN NEW ENGLAND'S VAUNTED TEXTILE industry could not keep up with wartime demand for warm sweaters, scarves, leggings, socks, and whatever else America's burgeoning armed forces needed. The government called on untapped womanpower for the solution. Women responded with zest.

BLIND CHAMPION OF KNITTERS

This Woman, Though Blind, May Be the Champion of the Army of War Knitters

By Marguerite Martyn

November 21, 1917. Mrs. Thomas A. Bruce of Salem, Missouri, has knitted fifteen sweaters, nine mufflers, five helmets and uncounted wristlets. The Red Cross has accepted them as standard.

And Mrs. Bruce is blind!

Mrs. Bruce is remarkable, for she also conducts an extensive business enterprise in her home.

You may have read in the *Post-Dispatch* Sunday Magazine

the account of this wonderful blind woman magnate, the owner and operator of the only telephone system in her part of the State. She installed the exchange in the face of persistent discouragement by well-meaning advisers, but she carried it to success.

Fourteen years ago, Mrs. Bruce was a dressmaker. It was not necessary, according to the standards of the little town, that any wife should engage in business. She might have been satisfied with the living provided by her husband from his earnings from a small real-estate office and the clerkship in a circuit court.

But she was ambitious to build a home beyond these means in accordance with her own taste and fancy. They had completed the house and were well established in Salem when blindness came.

Mrs. Bruce is not one to adopt a middle course. When she faced a revolution in her life, she chose no compromise but undertook a more ambitious venture than she had ever dared.

With the promise of the subscribership of sixty of her helpful but wary friends, and with borrowed capital, she installed a telephone exchange in her own home.

From that small start, the exchange has grown until now it serves more than four hundred local subscribers and forty country toll lines. It connects with Bell long-distance service, and it occupies a building of its own with a plant value of some twenty thousand dollars.

Staff of Eight

By day Mrs. Bruce superintends four operators. At night, she sleeps nearby and attends to the calls herself. She has eight operators, linemen, and truck drivers who look to her for leadership.

This past week, I met her for luncheon in St. Louis. She is an attractive woman with a glowingly fresh complexion and white hair. Her carriage is erect and confident. At table she keeps up a witty flow of conversation.

Through devoted women who read aloud to her, she has kept in touch with the news of the day, and when the war came, she wondered how she could serve the country.

She heard how soldiers and sailors needed knitwear. She ordered the pamphlets which the Navy League produced and had a friend read them.

Mrs. Bruce seems destined to excel in her new endeavor. Besides her own work, she has finished several pieces that her friends had begun.

"I feel that the faster I knit, the sooner the war will be over," she told me.

It would certainly be disheartening to the Kaiser if he could see the way her fingers fly through the wool.

• • •

Next, two women sit in an (oddly) empty streetcar, each busy with a knitting project. The chic one on the left is knitting something tiny with an enormous pair of needles; she on the right is producing what looks like a full-size comforter with a tiny pair, taken from the small purse at her side. (November 24, 1917.)

This shows that —

The Size of the Bag Does Not Always Count

PUNISHMENT: LEAVE KNITTING AT HOME

December 8, 1917. When Teacher has to punish the worst boy in school for not having his lessons learned, how does she do it these days?

No, she doesn't stand him in the corner with a dunce cap on his head. She doesn't seek out vulnerable parts of his anatomy with a rattan stick. She doesn't send him for a séance on the principal's carpet. No. She invokes a system more modern and improved.

She *forbids* him to bring his knitting to school.

This method of discipline is just one more sign of how far this fascination for knitting has brought us.

Last Saturday, three shabby, grimy little urchins were sitting unabashed and oblivious amid a group of lady knitters at Red Cross headquarters. They were struggling with needles and twine that had once been white, which beginners are given for practice. *(Next drawing.)*

AT THE RED CROSS KNITTING ROOM SATURDAY AFTERNOON

"Isn't that rather unusual?" I asked.

"Oh, no," replied Miss Dunnica, the instructor at Hodgen School. "Most children learn to knit at school from their teachers or from each other, but we often have them here on Saturdays for advice."

A little boy explained: "I don't have time to stay after school. I have to get out and sell my papers."

A hot trail had led me to Hodgen, where the principal, Mr. Christy, man-like, was at first inclined to treat the subject as something trivial. But soon he ceased protesting and was showing me a bulletin board where lists were headed "Sweaters," "Wristlets," "Scarves" and "Helmets," and children's names were inscribed as on a roll of honor.

These were the pupils whose work had been accepted by the Red Cross, and boys' names came close to outnumbering those of girls.

Energy and Enthusiasm

Miss *(no first name)* Stout said:

"I saw where a great deal of energy and enthusiasm and good wool, as well as patriotic service, might be used for our men at the front." She obtained the cooperation of the school patrons' association and a loan of $260 to pay deposits required by the Red Cross, "and we became a regular knitting branch."

To qualify for the Hodgen School R.C.K. (Red Cross Knitters) Association, a child must furnish an acceptable sample of plain knitting and purling. The child is then given an inscribed button that enables entry into the building as soon as the janitor opens it in the morning.

The privilege has worked wonders in curing tardiness.

"We never have to encourage the children to knit," Miss Dunnica said. "The day after the first lesson, one little girl brought in a sample she had done with hat pins because she had no needles. And a boy who began with a pair of meat skewers is now a leading knitter, much sought for advice and instruction in his room.

"I don't like to admit it, when knitting has so long been a feminine prerogative, but it is a fact that in speed, determination, and devotion the boys often outdo the girls.

"The enthusiasm is so great that we sometimes have to suppress it. I use it as a measure of discipline — if a pupil comes to class with his lesson unprepared, I forbid him to bring his knitting to school. That proves to be a corrective every time."

"But do sure-enough, rough-shod, two-fisted boys do this?" I asked, inspecting a beautifully smooth sweater.

"Come forward, Roy, this is your work," was Miss Dunnica's response. And a tall, red-haired lad, just the type I had in mind, stepped forth. He was awkward and shy, and his voice needed filling, but he made it known that his name was on the honor roll four times, with two pairs of wristlets and two sweaters to his credit.

"The boys draw the line, though, at carrying a knitting bag," Miss Dunnica said. "At first, they would buy large paper sacks from the grocer to carry their ball of yarn, but now I notice they resort to any sort of scheme to camouflage their new pursuit."

Martyn thought about this, then decided to illustrate it: A boy scoffs at the girl's flowered knitting bag, but the lad behind him discovers an end of yarn in the shoebox where the first fella had tried to hide it.

. . .

*But not everybody thought that knitting would help win the war —
or even that knitters were particularly patriotic.*

Mary Harris Jones, known as "Mother Jones," was a fiery labor leader once called "the most dangerous woman in America." Labor union leaders brought her to St. Louis to beef up the morale of strikers at the Wagner Electric Company — and to give them hell, too, because that's the kind of language she used. Although no newspaper would print her spicy verbiage verbatim.

WOMEN SHOULD BE PROTECTED, NOT VOTE

May 13, 1918. I cannot give to Mother Jones's words the eloquence they have as they roll off her tongue with their droll Irish twist amid the deep, sonorous cadences of a voice that belie her age, eighty years old last August.

MOTHER JONES — SKETCHED BY

Her expletives and swear words are far-famed, and she doesn't hesitate to use them in private conversation.

But you don't get an impression of hardness from Mother Jones. I found her just coming in from a shopping expedition with some buttons and pins she required to finish a black poplin skirt which lay spread out on the bed in the Majestic Hotel.

I sat meekly silent while her scorching tongue excoriated many institutions I myself had looked upon with toleration.

1. **She had little patience for women supplanting men in war industries.**

"I see them climbing all over engines with their oil cans. I see them pumping levers on streetcars; I see them pushing heavy trucks of munitions, and I think 'What of the future generation?'

"Women's nervous organism is not equal to such work. A principle of trade unionism is that women shall work under conditions that will safeguard to the utmost their bodily welfare."

2. She dismissed women's suffrage with equal scorn.

"Women vote in Colorado, and what have they done to improve industrial conditions? After the riots at Trinidad, where twenty women and children were laid out in the morgue, committees of ladies came to look over the scene, and they dismissed it, with 'Too bad, too bad!'"

The Trinidad, Colorado, event is known as the Ludlow Massacre, and the place is marked as a National Historic Site. The State militia killed eleven children and nine adults there during a lengthy coal miners' strike in 1914.

3. Women in war-relief work were beneath her contempt.

"They *must* do everything in public. If they knit, they sit out on their balcony or in some more public place, so they can be seen. They knit, so" — and she crooked her fingers in imitation of the painful awkwardness of a beginning knitter — "the while berating their maids and their laundresses and their dressmakers."

Spoken from experience perhaps, because Jones had once earned her living as a dressmaker. Yet she had not worked for wages; she owned her own shop that the Great Chicago Fire destroyed in 1871.

4. She condemned social workers, too.

"Social settlements, rescue homes, and organized charities are mostly supported by capital wrung from the blood and souls of those who have become the objects of charity and the inmates of these asylums. They are just part of a capitalistic system."

5. Women are kept in ignorance

She said of women:

"They aren't wholly to blame for their mistakes. Women's training is part of the capitalistic system. They are kept in ignorance. The only way they can learn is by experience in industry. *(That is, take a job in a factory.)* Why, if all women knew conditions as they exist in industry, they could revolutionize the world. If they wanted to."

• • •

Next, women trained in the capitalistic system and kept in ignorance decide to support the war effort anyway.

20. Blue Gingham Aprons

IN ST. LOUIS, THE YOUNG Men's Christian Association opened a new canteen for servicemen at 20th and Eugenia streets near Union Station, the central transfer point for troop trains of five states.

The structure cost ten thousand of the million dollars that St. Louisans had contributed to the Y.M.C.A.'s international war work fund. Women volunteers served as "waitresses and spirit-cheerers."

There was only one sign on the walls: "When did you write home last?" *That did the trick. In February 1918, the 34,257 men who used the facility left 30,267 pieces of mail. (Post-Dispatch, April 11, 1918.)*

Marguerite Martyn herself went down to volunteer.

TENDER MOMENT AT THE CANTEEN

By Marguerite Martyn

May 12, 1918. "Good-bye, lady."

The boy, still swallowing a last bite of apple pie à la mode, rose from his seat at the Railroad Y.M.C.A canteen and held out his hand, which I grasped. And he was gone.

Something in my throat warned me that in another instant I should be a "sob sister" sure enough.

But my soldier had already melted into the lines of khaki headed toward the platforms, and I joined the other women in doorways who were waving napkins and aprons to the hearty cheers of the men being marched away.

WHEN 600 HUNGRY MEN
ARRIVE ALL AT ONCE

This trainload of troops had stopped in St. Louis after three days of travel eastward, toward the war. They had a refreshing bath, then feasted on amazingly large helpings of food for checks purchased at five cents each.

Canteen Is Open From 9 to 9

Mrs. Charles Cummings Collins is head of the group of seventy women who keep the canteen open from 9 to 9.

I had arrived, not at 9 a.m. but in time to hear the announcement that a troop train of six hundred was unloading in the yards. Get ready for a stampede, we were told.

Five cents purchases any dish on the bill of fare, and the helpings are so bountiful that the women have to warn the men not to buy too many checks.

Waitresses who had been chatting sprang to their posts, some of them a little warily because there had been a false alert before. Great stacks of sandwiches had been made ready, coffee urns filled, and everything cleared for action. Then, with no explanation, the marching column had advanced to the corner, halted, then marched back to the train yards.

This time, though, the hall became flooded with men. They slammed their packs on the floor with a noise like artillery and set forth to bombard the lunch counter. Women in fresh blue gingham aprons and white organdy caps sprang into action.

Six hundred men were fed, and they ate. They were allotted about fifty minutes.

Then an officer appeared at the door and shouted orders to "Fall in!" With reluctance the men dragged themselves from their tables, gave a hitch to their belts and took a last look at the cozy interior of the canteen.

What sudden impulse had overcome an all-too-apparent shyness and made that soldier turn demonstratively to the first woman at hand?

Something in the manner of the waitresses, their amateurishness, their eagerness and good humor, perhaps?

Possibly it came over the youth that I represented a small atom of a big world he was leaving behind — simply an American woman. So he felt moved to grasp my hand.

So I am able to testify to the effectiveness of the "personal touch" which the organization aims to provide along with service of good, wholesome food at bare cost.

Often there is time for a waitress to hear the whole life history of a customer.

WHEN THE RUSH IS OVER
THE WAITRESS FINDS TIME
TO HEAR THE WHOLE LIFE HISTORY
OF HER COSTOMER.

Some men want to leave tips. One offered fifty cents to his waitress, who turned out to be Mrs. Collins herself. She promptly redirected her soldier's goodwill.

"The next time you meet a man who hasn't any money to eat, just give that half dollar to him — in remembrance of this canteen."

The soldier said he would do it.

• • •

In the next chapter, Clair Kenamore follows the troops.

21. First Days in Paris

CLAIR KENAMORE WAS STILL COVERING the Mexican border, when —

REPORTER GETS A NEW POSTING

By Clair Kenamore

April 25, 1918. I was in a town in the American Southwest when I received telegraphic instructions to proceed to Europe. I passed hurriedly through St. Louis, pausing only to make application for a passport. *(On Tuesday, February 18, in a St. Louis District Court, and I'll bet that Marguerite Martyn took some time off to be right there with him.)*

The official inquired as to my race and nationality.

"Was your father of German birth?"

"No."

"Your grandfather?"

"Nope. No German blood whatever."

"Very well."

He issued the application *(next image)* and signed it. His name was Otto Fickenstein. *(Which meant that, in the nationalistic fervor of the times, the clerk was of "German blood.")*

Thence to Washington, where an overcrowded Bureau of Citizenship tried to hurry my passport through — and failed.

On the way, our long passenger train lay on a siding while longer trains of empty boxcars passed westward. Passengers grumbled, but everybody knew that essential freight for national defense was more important. And miles of laden coal cars traveled the other direction, with occasionally a trainload of military equipment.

The application must be in duplicate and accompanied by three unmounted photographs of the applicant, not larger than three by three inches in size, one of which is to be affixed to the passport by the Department; the other two must be attached to the application and its duplicate, respectively. The photographs must be on thin paper and should have a light background. The one not attached to the application should be signed by the applicant across its face, so as not to obscure the features.

This blank must be completely filled out. The legal fee of one dollar, in currency or postal money order, must accompany the application.
A woman's application must state whether she is married or not, and a married woman must state whether her husband is a native citizen.
The rules should be carefully read before mailing the application to the Department of State, Bureau of Citizenship, Washington, D. C.

[Edition of 1917.]

[FORM FOR NATIVE CITIZEN.]

Issued

UNITED STATES OF AMERICA.

State of .... Missouri ....

County of .. City of St. Louis .... } ss:

I, ... Clair Kenamore ...., a Native and Loyal Citizen of the United States, hereby apply to the Department of State, at Washington, for a passport.

I solemnly swear that I was born at ... Eminence, Mo. ...., in the State of .... on or about the .. 22nd day of October 1875 ., 1.......;* that my {father} ... George R. Kenamore ......, was born in .. Columbia, Tenn. ., and is now residing at .. St. Louis Mo. ..
†[that he emigrated to the United States from the port of .... on or about ...., 1.......; that he resided .... years, uninterruptedly, in the United States, from 1...... to 1......, at ....; that he was naturalized as a citizen of the United States before the .... Court of ...., at ....

on ...., 1......., as shown by the accompanying Certificate of Naturalization]; that I have resided outside the United States at the following places for the following periods: ... Mexico ...., from March 1916 .. to May 1916 ...., from .... to ....

and that I am domiciled in the United States, my permanent residence being at .. St. Louis ., in the State of ... Mo. ., where I follow the occupation of .. Newspaper Writer ....
My last passport was obtained from ...., on .... and was .... (Disposition of) Indefinite .... I am about to go abroad temporarily; and I intend to return to the United States within {months}{years} with the purpose of residing and performing the duties of citizenship therein; and I desire a passport for use in visiting the countries hereinafter named for the following purpose:

.... France .... Newspaper Work ....
(Name of country.)   (Object of visit.)

....
(Name of country.)   (Object of visit.)

....
(Name of country.)   (Object of visit.)

I intend to leave the United States from the port of .. New York City ..
(Port of departure.)

sailing on board the ...., on .. March 1st 1918 ., 191..
(Name of vessel.)   (Date of departure.)

OATH OF ALLEGIANCE.

Further, I do solemnly swear that I will support and defend the Constitution of the United States against all enemies, foreign and domestic; that I will bear true faith and allegiance to the same; and that I take this obligation freely, without any mental reservation or purpose of evasion: So help me God.

Clair Kenamore
(Signature of applicant.)

Sworn to before me this .. 19th ...... day

of .... February ...., 1918

[Seal of Court.]

W. W. Hale

Clerk of the U. S. Dist. Court at .. St. Louis, Mo. .

* A person born in the United States should submit with his application an affidavit of the physician who officiated at his birth or of some other person who has actual knowledge of such birth, if the fact is not officially recorded, or an affidavit from the attending physician, parents, or other persons having actual knowledge of the birth.
† If the applicant's father was born in this country, lines should be drawn through the blanks in brackets.
[OVER.]

Washington was full of officers, majors seeming to predominate. Too many young lieutenants were on hand to suit my taste. They should have been abroad, leaving the desk jobs to the gray heads.

The War Department resounded with grim energy, with its guarded entrance and a pass demanded for simple entry. In the hotels the talk was all of war, and at the National Press Club likewise.

I got my passport, a stately thing signed by Secretary of State Robert Lansing, and I hurried to New York just in time to miss the boat.

• • •

New York seems to be aware that a war is going on but is not particularly interested. The people think everything done in Washington is done wrong and would be done better in New York. Sedition? No, it amounts only to that patronizing, superior New York attitude.

This is an air that has driven many a Westerner to vow in his first year in Manhattan that he will go back promptly and permanently to his own land. Eventually, though, the Western man becomes a New Yorker, and he takes on that same lofty attitude himself.

• • •

Finally we sailed away, on a ship which I cannot name, bound for a port which I cannot name. In our cosmopolitan company were:
- American soldiers going to join their units.
- Sailors trained as airplane mechanics.
- French and Italian officers who had been busy buying supplies.
- Lads from universities on their way to help out in consulates.
- Women doctors in the uniform of the Red Cross.
- Younger women who had volunteered to work in canteens and had paid for their own tickets.
- A frail girl from the West, reputed to be very rich, who would drive a "camion" *(truck),* which she herself would pay for and keep supplied with gasoline.

- A dozen Salvation Army girls who were to sew on buttons and darn socks for soldiers.
- Two men from the Japanese Y.M.C.A. who restlessly walked the deck in all weather, bareheaded. *(Japan, intent on seizing German colonies in the Pacific, had joined the war.)*
- Famous opera singers and a woman who broke the rules against smoking on deck after dark, the glow of a cigarette a tipoff to enemy submarines. *(Almost forty ships were struck by German torpedoes during 1917.)*
- A blind soldier who had learned how to wind armatures and was on his way to France to teach other blind soldiers.
- A fat Frenchman with a fierce mustache. He was called the "sea wolf" because of his persistent gallantries toward the Red Cross nurses.

Better Than a Concert

There was the little Belgian who had been ten years a shoe clerk in Chicago but was returning to fight for the wisp of his country that remained free *(the Germans having conquered all the rest).*

He was the life of the ship. An afternoon on deck with him was far better than a concert in the main salon with Dalmorès and Dufranne. *(Charles Dalmorès was a French operatic tenor and Hector Dufranne was a Belgian bass-baritone.)*

He had forty-two days of leave after three years of hell, but in the gayest voice, he would sing "Joan of Arc, They Are Calling You" and "America, Here's My Boy," and finish up with "Barney Google With the Goo-Goo-Googly Eyes"!

He helped out with the French and Italian singing bees and ended up dancing with the gray-haired lady doctor.

• • •

When the ship entered the danger zone, our stern guns were slewed around to cover the seas to broadside, and the bow gun was aimed straight ahead. Sailors climbed to the crow's nest and kept a constant watch.

There had been no deck lights from the start, but now the most stringent rules were in effect.

Flashlights were not to be carried, matches had to remain unlighted, smoking was forbidden, even behind the canvas curtains. All portholes were blinded with steel plates, and no light showed from the great vessel.

Dark and blind, she plunged through the night at her best speed. It was better to take a chance on colliding with another blind ship than to show a gleam to an enemy submarine.

The last day and night, we zigzagged like a drunkard, and then, finally, came France.

• • •

We were in the war. The streets were almost bare of men. Outside the railroad station, women waited for their loved ones to come home on leave. Sturdy, round soldiers they were, with rifles, heavy packs, and helmets slung behind them.

Some men did not arrive, and some women wept. Others comforted them and spoke of "wounds" and "hospitals."

That night, I was in Paris. *(The 35th Division, which Kenamore was chasing, was at Eu, Seine-Maritime, on the English Channel, far behind the lines.)* Enemy airplanes were overhead. The musical rumble of anti-aircraft guns. The huddled crowds in the air-raid shelters.

Overnight, a shell from a long-range gun struck near where I slept. People were killed, I was informed the next morning when I saw the place. It was war. I had come part of the way seeking it. Fritz *(the German invader)* had obligingly sent it to meet me.

GERMAN GUN DAMAGES PARIS CHURCH

April 23, 1918. On Good Friday, the day of the Crucifixion, and at the very hour the Savior died, the German long-range gun resumed its bombardment of Paris. *(This huge howitzer weapon seventy-five miles away had several nicknames, including Lange Max and Big Bertha.)*

Wednesday and Thursday it had been silent, a respite that gave rise to many stories. The one most accepted was that our aviators had discovered the forest where the gun lay hidden. Relays of fliers then soaked this with poison gas and saturated it with inflammable oils. They then dropped incendiary bombs, and the whole wood burned. *(None of this was true.)*

Parisians were not at all panicky, but they were relieved that the cannon was silent. Many left the city for the south of France, yet there was no way of knowing which was the holiday-goer and which was the person whose nerve had failed.

Then came Good Friday, March 29. At the "ninth hour," when Jesus expired on the Cross and which was 3 p.m. local time, congregations were in churches to mark the event.

A single church in Paris was struck.

To believe that the gunner selected that edifice does too much violence to human credulity. To hit a mark no bigger than a building from a distance of seventy-five miles cannot be the result of skill. It must have been the Devil's own doing. One shell filled with explosives landed within this sanctuary.

The church was just one street east of the Paris City Hall, so the aim of the German gun on this particular day was pretty accurate.

I cannot name the edifice or its *arrondissement* because that knowledge would help the enemy. The church was not great and famous, but it was known for the dim beauty of the interior and, a priest tells me, for the piety of those who worshipped there. *(It was the Eglise Sainte-Gervais. You can still see the shrapnel marks in the walls of the nave.)*

It had a vaulted roof supported by stone columns. The shell tore its way into the building and did not explode until it struck a support, high above the worshippers' heads. The pillar was destroyed, masonry fell upon the people below, the roof sagged and collapsed. *(Photo following, from the French National Library.)*

This triumph of German ingenuity and valor took the lives of eighty, and ninety-two were wounded. Most of the victims were women and children.

The object of this bombardment of Paris can only be the moral effect on the civil population. The Germans send shells in the hope they will terrorize the Parisians. It will aid the work of the Defeatists, a hidden band of German propagandists who argue that the Boche cannot be defeated.

So far, the big gun has failed.

When the attackers were in airplanes, Parisians felt a sporting element because defenders shot down a good number of them. But the new weapon has brought only anger, resentment, and determination. That men seventy-five miles away would strive in comparative security to send shells into a city peopled largely by women and children — it was too much! It was too essentially German. It was too un-French.

22. Journalists and the Armies

WHILE MARGUERITE MARTYN continued her interviews and sketches on the home front, Claire Kenamore was learning how reporters covered the war in Europe. After the war, he recalled:

JOURNALISM POLICIES DIFFERED

By Clair Kenamore

May 11, 1919. Each of the armies in the Great War had a different way of dealing with war correspondents.

The French lost a battle to the Germans in the War of 1870 because a London newspaper printed a story showing the direction one of their armies was moving. Paris, in its rigorous censorship, saw that such a thing would not happen again.

As for the British, Sir Garnet Wolseley once said that all war correspondents ought to be hanged. *(During the 1873 Crimean War the bemedaled field marshal complained about journalists, "those newly invented curses to armies, who eat the rations of fighting men and do no work at all.")*

No serious person should question the need for censorship. The arguments were about interpretations of the rules, and over them the correspondent and the censor went to the mat.

The British journalists had a beautiful old chateau in a park at Rollancourt in the north of France *(where Kenamore first stationed himself)*. From there each day they went to whatever part of the front they chose. But they were not allowed to move around without a conducting officer.

Much of the news came in via telephone. Frequently, the reporters from the press associations, called "flash" men, never left during the day because they could do the reporting better from there. *"Flash" was the term used as the header of a brief message that a major story had broken and details would follow.*

These men did go to the front — for example, when they wanted to get an idea of the intensity of artillery fire, which had to be seen or heard to be comprehended.

Lack of Competition

At most any time at Rollancourt, I could see a studious Briton or two strolling about, deep in thought. They would be correspondents digesting the news of the day before writing their stories.

There was no competition among British reporters. When they returned from the day's work, they got together in the main hall and everybody laid his cards on the table: One man chose this item to lead his story, another took that, and each wrote his article. Each piece thus depended heavily on the style of writing for success.

The war was won and lost many times over via the discussions around that tea table. I hate to think what would have happened to the tea cups if the Americans had had such a policy.

As for the French, their correspondents were allowed at the front only when the army thought best, and they had less access to the official reports, especially the confidential ones.

The Americans were more liberal. Our army keenly appreciates the great need of boosting morale back home — by giving the people news about the boys in the field.

Not Suited as a Censor

So our press section was established under the direction of Colonel Dennis Nolan, chief of the Intelligence Department of the American Expeditionary Force.

That press section was known as G-2-D, headed at first by Major Frederick Palmer — but he was soon transferred to the Historical Section after he declared himself utterly unfit for the job of censoring, having played on the other side too long as a correspondent. Major A.J. James then took over.

Palmer's first war as a journalist was between Greece and Turkey in 1897, and his last foreign assignment was atomic bomb testing off Bikini Atoll in 1946.

Army headquarters were in Chaumont, near the sector the Americans were to occupy, but the press section was in Neufchateau, some thirty-five miles away. A branch was established in Paris at 10 Rue St. Anne, and censors were at both places. *(There's a fire station there now.)*

The Americans gave correspondents greater freedom. Associated Press and United Press had men in Paris to keep in touch with government offices and to get what news they could from the Paris papers. In late spring and early summer of 1918, when we did not have many troops in France and the lines were held by the trench system, the other American journalists gathered at Neufchateau.

The photograph shows Kenamore at far left, and behind him is J.W. Grigg of the New York World. The last man in the front row is John T. Parkerson of the Associated Press. The others are unidentified.

With some exceptions, reporters with our troops were divided into accredited and visiting correspondents. I was accredited, so I took an oath just as officers did and was, in fact, part of the U.S. Army. We had passes which let us go through all lines.

The Army provided automobiles, with the upkeep divided among the accredited correspondents. The other reporters used them as well. Daily the motor cars left Neufchateau for the front. Each evening they returned, and the press association men filed their stories. Some of us stayed overnight in the trenches or billets occupied by our troops.

The *Post-Dispatch* was most interested in the 35th Division, the Missouri-Kansas men, in the Vosges Mountains. To get there, I started from Neufchateau very early in the morning and drove about fourteen hours, arriving in the high hills after dark.

From Wesserling, or Larchey, or wherever I stopped, I sent the automobile back to Neufchateau and stayed with the troops until I had gathered enough material and pictures.

Sometimes I wrote the stuff in the Vosges, sometimes I went back to write it. I usually had to walk down the mountains to Kruth or Wesserling, a day's work if I couldn't get an empty ambulance. Then I'd catch a military train, maybe a freight, for Epinal or some place on the main line.

Argument With the Censor

Thence I could hop a train to Nancy or Chaumont, and from there a truck or car to Neufchateau, where I'd hand over my work to the censor.

There would be considerable conversation and argument because we never seemed to agree how nearly I could identify a particular unit, which of course would be of great importance to our readers.

Once the story, more or less mutilated, had the censor's red stamp on it, the telegraph operators could send a limited amount, and the rest could go to the French post office.

Then I would catch a train for Paris to mail off my pictures.

I had to give all film to the Army Signal Corps to develop and print. The only lab able to do that work was in Vincennes, seven miles from the capital. It took three or four days to get the prints, and then I had to take them to the picture censor, and he kept the ones he didn't want to go out. I wrote captions on the others, had those approved, and mailed them to the *Post-Dispatch*.

Then it was back to the front and through the same process over again.

Enough News for All

After the Americans had sufficient troops to take over a portion of the line ourselves, it got simpler. The Army established field censorship headquarters — first at Beauvis, Meaux, and Nancy. For the Argonne fighting and all preparations, we were at Bar-le-duc, in the Meuse, for about six weeks. Finally, Verdun and Coblenz.

Each of the press associations had at least two men in Bar-le-Duc, and some English correspondents joined us. There was an abundance of news for everyone, especially the flash men.

The American aviation camps were nearby and always good for a story, and there were the various corps headquarters visited once or twice daily by the association people at least.

Our troops held a stretch of trench at Toul, and some others were in line with the French, learning how to whip the Germans. Soon we knew that we were going to "pull a big show."

Plans for a Big Battle

One afternoon Major James appeared from Chaumont and said that General Dennis Nolan wanted to see everyone at 8 o'clock. We all went, and the general gave us a dinner.

Each correspondent received a set of five battle maps which were duplicates of those pinned to the walls. Then the general spoke for about ninety minutes. The American Army would attack at 5:30 the next morning, with fifteen divisions, between the Argonne and the Meuse. It was to be the most powerful army the United States had ever sent into the field. He described the action in detail, giving minute plans to reduce each German strong point, the assault on each height, and the whole plan.

By the time we had finished asking questions, it was 9:30 p.m., and the artillery preparation for the advance would begin at 11.

Then began a great scramble for automobiles. We crowded into our machines and started northward. It was dark, and motor transports of all kinds were plunging through Bar-le-Duc. We could have no lights because the Germans had bombed the town from the air, and there had been an "alert" against enemy aircraft almost every night.

The next image was distributed by the U.S. Committee on Public Information in September 1918 and appeared with the same headline in newspapers all over the country. These are the autos ready to head out.

WAR CORRESPONDENTS TAKE DARING CHANCES

The Voie Sacrée, the Sacred Road, was alive with traffic. Trucks carried shells, rations, forage, kitchens, ammunitions, and men — the reserves taking up their positions.

This marvelous event could transpire only in wartime: not a light on any of the thousands of machines, but constant movement at a fair rate, about 15 to 20 miles an hour.

One military policeman loomed from the darkness and screamed: "Stop that truck! Stop that (blank blank) truck!"

The *camion* stopped with its radiator only an inch from the tailgate of the truck in front. The rear guard of the front truck said calmly:

"Yes, better stop it. This one is loaded with hand grenades."

Bombardment Strikes the Enemy

The early bombardment at each end of the American line began at 11 p.m. The great hammering of twenty-six hundred cannon was to follow at 2:30 a.m.

By 2 o'clock the traffic was so sluggish that we drove our machine off the road and pressed ahead on foot to see the artillery begin its work. We found a good place on a hill, behind the infantry, and waited. *(Infantry attacks were always preceded by heavy bombardment to soften up the defenders. The blind shelling from miles behind the front line proceeded just ahead of what was calculated would be the soldiers' advancing position.)*

Our timing was excellent, and the big guns opened with a roar, and for three hours they kept booming with the speed of machine guns. We lay there while these drums of doom rattled away; then they ceased and reset their aim to deliver a deadly barrage just in front of our infantry.

Among the artillerymen was Captain Harry S. Truman, who twenty-six years later would become President of the United States.

We cheered the doughboys as they went forward, then we separated to find our various divisions. I got to the 35th by noon and got a whale of a story.

The division was advancing in the valley of the Aire. That first day was favorable. The Prussian Guard, the best in the Kaiser's army, was there to meet the Missourians and the Kansans, but our men were whipping them in the open field.

Tail End of a Motorcycle

I had to get to Bar-le-Duc and move my story to the *Post-Dispatch* office, some five thousand miles away. I must have ridden on twenty different vehicles on my way, and I arrived soon after dark on the tail end of a motorcycle.

Most of the fellows were back, and they had already held a lottery to determine in what order they might file. The news agencies had persuaded the head of the section to let them go first, with two hundred words apiece.

Then came us specials. Everybody pounded out his piece, took it to the censor, quarreled with him, threatened to tear up the story altogether, decided not to, took what was left, and hurried it to the post office, where it was telegraphed back home.

We got something to eat, usually at the French officers' mess, found a car, and soon we were pounding northward again.

The first week of that battle, the correspondents did most of their sleeping in automobiles.

After the first day, the traffic was better handled, and we could get near the fighting by auto, then walk forward to a brigade or division headquarters, find out how things were going and what had happened since yesterday, hustle back to the machine, and be driven to the telegraph and the censor.

You carried food with you and you got coffee from anywhere. Sometimes you had hot beef and beans, but you relied on corned beef and canned salmon, which the soldiers called corned willie and goldfish.

Long-Lasting Traffic Jams

A traffic jam might last from a dozen to twenty hours, and it always seemed to be a road you could not escape because it would be hugging the edge of a cliff. You could get out and walk for five or six miles, or you could just wait and sleep.

But congested roadways are a tempting mark for enemy aviators or artillery, so you were often awakened by an alarm or the whistle and boom of an aerial torpedo. Then you climbed out and got a hundred yards away from the road, sat, and waited. Occasionally some ambitious German bird would strafe the jammed traffic with a machine gun; that added diversion.

Our news went across the ocean, day by day and night by night. Not only the battle communiqués, but also the tales of the men themselves, the soldiers, the wounded, the couriers, and the commanders.

New ways of warfare

The aviation and other new ways of warfare produced lively picture stories, and the trains, the liaison, intelligence, propaganda services did good and essential work. But the infantry and machine gunners were the forces that made the news, big and decisive, right where they were at the moment. It's the infantryman with a rifle who decides the issue in war and the value of the front-page yarn alike.

In handling important news like the start of the Argonne offensive, the press officers determined the capacities of the wires out of Bar-le-duc and got the Army to allot the press a certain amount of traffic. Say they were allotted ten thousand words a day — they would then put the names of the correspondents in a hat and draw them out one by one. The reporters could file a hundred words each, in the order drawn.

There were some twenty cable men there. Most of the early stuff was filed "URGENT," which cost about 75 cents a word. After that first batch, you could file three hundred words.

The censors were Army officers who had been newspaper men, and I found them honest and conscientious. I differed with them on judgment, advisability, propriety, or some rule handed down by the Grand Headquarters, and they were a hard-headed bunch.

They were doing their share in the great common cause, just the same as the soldiers in the fields, but they spoiled a lot of good copy.

There were also some very likable French censors, and they seldom asked us to change a word.

Correspondents wasted time running down rumors. There was the good old fake of the German woman machine gunner found chained to a gun; that story cropped up after every advance. Reporters went for miles trying to find somebody who had seen such a person. No luck.

I was primarily interested in the 138th Infantry Regiment, mostly St. Louis soldiers. Once I stopped at division headquarters, and the chief of staff took me to task for the alleged delinquencies of "my" regiment. He said the men were robbing beehives and wine cellars.

"What sort of outfit is that, anyway?" the officer demanded. "Just look at this bunch of claims made by civilians. They must have professional burglars in that outfit. Go down there and find something about them, see if they plan to crack any safes."

I responded with great hauteur: "Those charges are libelous. Men of that regiment do not know how to rob beehives. That must be some other state outfit, probably from Kansas. The 138th is filled with fine, high-spirited moral Missourians."

I left him and went to join my boys. I couldn't get a billet and slept in the open, as it was a pleasant summer night. In the morning when I awoke I found somebody had stolen my shoes. I had to ride a mule down the hill to the supply company to draw a pair of field boots.

23. Marriage, Yes or No?

"SLACKER" WAS A BAD WORD in 1917, as we saw in Chapter 8. Martyn's cartoon below (August 9, 1917) has a military-garbed Cupid (except for the bare feet) holding up a tiny palm to the dippy-looking couple on the City Hall steps.

"Halt!" Cupid demands. "Are you in love with each other? I want no slackers in my army either."

Inside the Marriage License Bureau a soldier has his arm around his fiancée as they stand at the counter. He at least is wearing shoes.

Cupid's Challenge to the Marrying Slacker! A Cartoon for Women By Marguerite Martyn

"Halt! Are You in Love With Each Other? I Want No 'Slackers' in My Army, Either"

The fourth Selective Service draft registration was held on September 12, 1918. For the first time, millions of older men (aged 30 to 45) had to sign up. Bands played, flags fluttered and patriotism ruled.

In St. Louis, workers for women's suffrage were outside the signup sites with petitions and the message "If you are going to fight, your wife or mother ought to have the right to vote."

Men who had a dependent wife could be exempted. That impelled Marguerite Martyn to draw this cartoon (September 5, 1918):

Will You Be a Depending Wife or a Defending Wife on September 12?

Two draft-age men have been asked the question, and little Mrs. Draft-Dodging Spouse crouches to hear her husband's fudgy response: "Yes, sir. My wife is absolutely dependent upon me."

Behind them, another man's wife, chin held high and as proud as a Girl Scout, declares: "I shall support myself in his absence. It is as much my country as his that we are defending."

No indication if any of them signed the suffragists' petition outside.

• • •

More than a million men were eventually sent to Europe. For the women they left behind, there was a problem — "When Shall He Marry? Before He Goes to War Or When He Comes Home?"

In Martyn's drawing below, which illustrates the article on the next page, a woman ponders her alternative futures.

On the cards: A bridal couple. A woman alone. A wife and a one-legged husband back from the war. A black-clad widow with a baby outside an employment office. A military officer with a suitcase. And a card with a bag of money and the word "Pension" on it.

When Shall She Marry? Before He Goes to War Or When He Comes Home?

What card will fate hold in store for her?

WHEN SHALL SHE MARRY?

By Marguerite Martyn

January 25, 1918. Should she marry her sweetheart before he goes to war, or wait until he returns?

"Yes, now! Don't delay!" That's the popular answer these days.

Marriages have increased to a startling degree last year, and it's the exception when the groom is *not* a soldier.

• • •

In a downtown girls' rest-recreation-lunch room, where everybody knows the others by their first names, Betty was showing off the circlet on the third finger of her left hand.

"Oh, I know," she chattered, turning her head before the mirror, bird-like, to see that the lipstick had done its work symmetrically. "I know it's fashionable, and all the society girls are doing it, but some of them will have lots of time to repent.

"Marry him before he goes to France? No! Why, he might come home a cripple, or leave me as a widow with a child to support. None of that for me."

And with a flattening pat of the wave in front of each ear and a backward glance at the mirror, Miss Betty sailed out.

"The mercenary wretch!" exclaimed little old sentimental Miss Quince, setting the teapot down with a bang.

"She ought to be thankful that a nice young man wants to marry her, the vain, silly-headed creature. Wait until we've sent our second million and our third million men to the front and there's no more left for her to choose from — and she is growing old, alone and uncared for — oh, she'll wish she had taken a fine fellow like Eddie Brown when she had a chance."

Marie, the little war bride who hadn't quit her job (hadn't even missed a day's work since she married a soldier last month), cuddled up to Miss Quince for an instant, almost tearful.

Susan, the sophisticated blonde who'd wed a sergeant and kept it a secret until he came home on leave, smiled in satisfaction at her reflection in the mirror.

Stylish French competition

Irene, who had been comparing engagement rings with Miss Betty, said:

"Well, I won't let my Billie go to France without we get married first. Think I'd turn him loose among all those stylish French girls? No! My only fear is he'll be hustled away with the next draft before I've got my trousseau bought!"

Frowsy Madge leaned forward in confidence.

"You'd better marry him the minute he's drafted. Then you'll have a lot more to spend, for they will hand you an allowance out of his pay, and if he gets killed, you'll get a pension and all that life insurance."

Irene answered this advice only with a withering look.

Anne spoke up.

"I've about decided to do it," she said. "You can all stop your knitting and begin tatting for my lingerie shower. When I think of turning Joe loose among all those gay French girls and sly Italians and wise English without a ring on his finger, I'm inclined to steer him over to the City Hall myself."

She frowned.

"But he made me promise I won't go out with or have anything to do with another man while he's away, and I'll die of loneliness."

A woman in the corner responded. She looked experienced beyond her years.

"It won't be any worse than having him here!"

The conversation stopped, and the woman explained:

"I never was lonesome in my life until I got married. He used to call on me every night of the week. But now that he's sure of me, he's picked up with his old gang at the lodge and his club and the corner drugstore!"

She Can't Make Up Her Mind

A silent little maiden, Jeanette, was gravely munching her luncheon. Now she spoke:

"You girls almost make me decide to do it, especially you cynical ones," she said. "As it is, I have to force myself from eloping with Bob every time I see him in his new uniform. And he almost persuades me by saying he is going right off and get killed if I don't! Still, when I am alone, I can't make up my mind."

Miss Quince said that Jeanette should do it.

"But he will see a lot of the world, and that may make him think I am more prudish and exacting than ever," Jeanette countered.

She sighed. "Still, he may meet so many of the other kind that he would appreciate a stay-at-home, commonplace girl like me."

She wavered again. "But my parents insist I shouldn't tie myself up for a future so uncertain."

Miss Quince snapped: "Your father and mother didn't worry about *their* parents. Didn't you tell me they had eloped?"

"But they don't want me left a widow with maybe a helpless little baby. Yet, they don't want me to be an old maid either. And," she added boldly, "I'd love to have a son to raise as a soldier for Uncle Sam!"

What will Jeanette decide? No doubt there are war brides who read this who can give her some advice. And there are other women with ideas to spare. Take your pen in hand and answer —

Should Jeanette marry her sweetheart before he goes to war or wait for that very indefinite coming-home?

• • •

Readers' responses were prompt.

ANSWERS TO JEANETTE'S PERPLEXING PROBLEM

January 28, 1918. (From Mrs. A.R.W.) "Girls want to do brave things and assume a share in the burden of the war. But women are subject to patriotic intoxication. I do not think that Jeanette's question is an emergency, like the draft of our soldiers. In her case discretion is the better part of valor.

"The movement to conserve the babies already here is more urgent than the births of more soldiers for Uncle Sam."

• • •

(From Mrs. M.O'C.) "I have two sons in the Army. I am glad neither of them is married. I think it a shame how the girls run after the soldiers. All they want is the pension or the allowance.

"This war is bad enough, taking young men away from their business just as they are getting started. As for the boys behaving themselves and keeping out of the way of scheming girls, every boy who has had the proper raising will be safe and will behave for the sake of a mother as much as for a wife."

• • •

(From Miss I.C.M.) "My dear Jeanette: Your uncertainty about your marriage has brought you very close to me, for I confronted the same problem. In answering it, I changed from a light-hearted, merry, happy girl into a sour, crabbed old woman, robbed of the best things of life, denied the joy which is every woman's heritage.

"In 1898 I was engaged to a young man who enlisted after President McKinley's call for volunteers. We wanted to marry before he left, but my mother would have none of it. With heavy hearts, we conceded to her and we promised to be loyal and true until his return.

"I am still loyally waiting. He was killed at the Battle of San Juan.

"There is no word to express what I felt at the news of his death. My mother became a stranger to me, for it was she who had robbed me of a woman's greatest happiness, wifehood and motherhood.

"I had promised to wait for him but daily I prayed to die. What would I not give to have heeded the dictates of my heart, to bear his dear name, perhaps as the mother of his child? How gladly I'd toil, how willingly sacrifice all in life for the realization of my dreams.

"That was twenty years ago. I no longer feel so bitter toward my mother and toward wars, but the sorrow seems as near, the regret as keen.

"All around me girls are asking the same question which we asked in 1898 and to all I give the same answer: 'Marry him.' Should he never return, are you not proud to be his widow, would you not be proud to be the mother of his child?

"Nothing in life can give you the same intense, heart-breaking sorrow you will surely have if you do not marry him and he never returns."

24. With the Americans in France

BRINGING BACK A PRISONER

By Clair Kenamore

May 5, 1918. *(Published June 30.)* The demand for a German prisoner became loud and insistent. None had been taken for a long time. Nobody knew what changes had been made in the enemy facing us.

Commanders demanded the identity of the other forces. Were they Austrian, Turk, or German? Were they old reservists (indicating an immediate future of quiet) or were they line regiments, or "circuses" of storm troops, which would foreshadow a charge on our lines or some other offensive action?

We call them "circuses" because the Germans move them from place to place, where they put on their perilous act of leading the way across No Man's Land. These desperate fellows jump from here to there in special trains and are not required to do the routine work of camp and trench. They are specialists who rest when there is nothing in their particular line to do. It is a gay life, and usually a short one.

Our soldiers also had the duty of reporting to Allied intelligence.

The ideal method of finding out what is up with those strange, hidden men in the yonder trenches is to go over and get a few of them, bring them back, and examine them.

This is how we did it, with fresh troops ready for the job.

• • •

Three hundred men, all dressed as if for a masquerade, were going over the top, not to return until they got their quarry. Through the wire and into the field, crawling on hands and knees, flattening for star shells and signal pistols, but pushing on.

At the first German trench they paused, cut the wire, filed through, and leaped into the ditch. It was deserted. They went ahead and found a well-protected dugout. A lieutenant from Wichita, Kansas, opened the flap of a blanket that served as a door and called, "Come out!" He did not heave down a couple of hand grenades, as was the custom, but he took his pistol in one hand and trench knife in the other and moved down the steps.

A sergeant aimed his electric torch and revealed seven Germans. Four held up their hands. The others were killed.

The squad hurried the prisoners into the trench, stripped the dugout of all papers that could be of value and headed fast for home, with the prisoners that the general staff had demanded.

The next day the Wichita lieutenant received a Croix de Guerre from a French general.

Kenamore's photo shows a prisoner being questioned. (Published February 10, 1919.)

MUSTARD GAS ON THE FRONT LINES

May 5, 1918. *(Published June 7.)* Mustard gas, which is the favorite frightfulness of the Hun, does not smell like mustard at all. Its pungency is something of the taste of mustard, but its smell is that of soured, fermented raspberry, with mold on the top.

The photo, taken by Kenamore and published April 4, 1919, shows wounded or gassed men at right being helped by other soldiers. At left, a mounted soldier leads a horse that has lost its rider.

Mustard gas is neither as refined nor as effective as other enemy gases, and it can be detected by the smell. It is heavy and clinging. Doctors in the clearing stations have been stricken by the stuff sticking to the clothing of the men brought in from the front.

A soldier relies on his chewing tobacco to detect the gas. If it suddenly goes tasteless and flat, he immediately sniffs the air, and if there is a suggestion of foreign odor, he yells, "Gas!" Smoke from pipe, cigarette, or cigar cannot be identified when there is gas in the air.

In the sector held by Americans, No Man's Land *(the shattered space between the two sets of trenches, American and German)* is too wide for "wave" gas attacks. Now the enemy is not using so much gas in the front lines, but rather cannon to drop gas on our artillery and our reserves in their billets.

Soldiers must then put on a gas mask, which reduces efficiency from 25 to 30 percent.

There are many embarrassing and harassing undertakings during a war. To target latrines with shells is some artillery observers' delight. Or kitchens just before mess time.

• • •

No man can fathom the mind of the artillery commanders. At 3 a.m. last night a battery of French 75s burst forth like hell boiling over. To have these previously unsuspected raucous war cats howling virtually under your bedroom window after you have returned from No Man's Land!

And it must have been good shooting, too, because Fritz replied with a few gas shells of his own dropped into town.

"Whan-you-whan-you-whan-you!" went the alarm, so every weary warrior had to turn onto his back and pull on his gas mask until the "All clear!" was given in twenty minutes and all could return to sleep.

SCENES FROM THE WESTERN FRONT

May 6, 1918. *(Published June 30.)* Dawn came, and we ventured soggily forth to see what kind of site had been chosen for the war. No Man's Land was a hollow field, filled with mist and rain. Barbed-wire entanglements began a hundred feet away and stretched farther, like noisome weeds.

As the wet, lifeless day advanced, we could make out the German line across a little, flat valley. It skirted the edge of a woods, then cut into the open for a bit before it wandered back and was lost in the trees.

The Hun was eight hundred to a thousand yards away, and here we were at last on the "frontier of freedom."

We were in a rotted trench, built three years before, when the retiring enemy had taken up his position across the valley. There had been no change since then; the Homeric battles for Verdun, a hundred miles to the north, had left this place unmoved.

The sides of these water-soaked trenches are held steady by revetments of willow branches, plaited like baskets. Upright posts, planted at frequent intervals and joined at the top by a connecting bar give the makeshift walls some rigidity.

The crossbar overhead connecting the posts are just short of six feet above the duckboards, thus placing them where the steel helmet clanks on each. The progress of a tall man passing through the trench sounds like a scissors-grinder coming up your street at home.

Tangles of barbed wire are placed on each side of the trench. Once the war is over, the French farmers will have enough of this wire to last them for years, and once the duckboard is dug out and the mud knocked off, they can heat their homes with it for a long time.

You can see the duckboard at the bottom of this trench, where soldiers lie asleep, or dead, and another keeps watch with his bayoneted rifle (Encyclopedia Britannica).

Our guide took us about a mile, where we came to a fork that led to an observation post. It is a round pillbox, with a slit an inch wide running around the side facing the enemy. The six men who hold it actually live in a dugout well to the rear.

The observer sits in a swinging seat, like the ones house painters use, and spends his two-hour watch scanning the field through binoculars.

The pillbox man keeps an accurate record which is sent back to the battalion intelligence officer every twelve hours. An example —

A.M. 9:11. Boche plane seen well back of his line.

A.M. 9:22. Two Boche wearing blue coats seen in edge of wood.

A.M. 9:25. Thin smoke seen to left of C14.

The observers thus identify every point by letters and figures. Back in headquarters the reports are plotted on a map. American and French reports call the enemy "Boche." British reports usually use "Hun." That is in the reports. It would not be polite to say what they call him in spoken words.

"German" is seldom, almost never, used.

Enemy Trench Is Concealed

The view is limited by the woods on the other side, which conceals the enemy trench. In the field just ahead I could see a grave, which appeared to have been there for many years. There was also a pile of rocks, broken up as if for road repairing, and shaped up symmetrically after the French fashion.

All these things were just as they were when the enemy retired across the field more than three years ago. Occasionally a shell had been lobbed over to maintain the status quo.

An intelligence officer came. A man on the line had seen a periscope come up from behind the rock pile and stay there for four full minutes.

The intelligence man was sure the enemy had sent someone out during the night to lie behind the rocks. That theory was supported by the fact that one of our men was shot when he injudiciously showed himself above the parapet.

Sniping and counter-sniping was part of the intelligence officer's work, and it was up to him to see that any adventurous Boche sent forth to get information did not return home. The officer thought it best to plant a sniper in the graveyard on the left to snipe the sniper.

We departed the post and went back through the dreary mile of disreputable trench. The big guns were languidly firing on both sides. One, two, three, four, they would go, then a long pause.

A Good View

It was agreed that at 11 o'clock our batteries would try to blow up a suspected machine-gun emplacement. I was planted where I would have a good view.

Our guns were three miles back, and the target was half a mile in front. To lie under dripping shrubbery in the rain and watch high-explosive shells fall into tangled woods is not an ideal way to spend a May day in France. The pleasure came from watching the marvelous accuracy of our guns.

They are far out of sight behind us and across some low hills and valleys, but they plastered their target with uncanny certainty. Brown clods of earth fountained up. The rain beat down the smoke and dust almost immediately, and there was no way to tell if a machine gun had been destroyed or a gunner killed.

• • •

We went back to town, which was quite new to eyes which had seen it only in last night's darkness. Four years ago it must have been a nice place, with about four thousand people. The front line ran through its eastern fringe, going right between two houses.

My afternoon was well spent in slumber in a child's deserted bedroom. She had liked Paris newspapers, and on the wall she had pasted many pictures from *La Vie Parisienne*.

Maybe even this one, from July 1914.

25. Four Sketches

MARGUERITE MARTYN OFTEN DREW THE shop girl, like the woman behind the counter of the department store who brought down the yardage from the top shelf or spread out the gloves before the seated customer. That working woman aroused her compassion.

One Letter Makes All Women Kin

March 21, 1918. *Work pauses and all await a reaction from the shop girl on the left as she reads a letter from her sweetheart. Even the usually officious floor manager is silent for once, realizing he is intruding on a solemn moment.*

• • •

HEARING NEWS FROM THE FRONT

April 10, 1918. *Two women have opposite reactions when they hear news from the front. She on the left is almost paralyzed as she digests the newspaper and holds a wadded-up napkin to her tear-stained face. At her elbow is a bottle labeled* "Nerve tonic."

On the right, a woman with a Red Cross emblem on her cap rolls up her sleeves for work; through the window she can hear a newsboy calling "Extra! Extra! All about the war!" *A colleague trundles in a new load of yardage that can be turned into bandages for the wounded. A scissors lies on the table, ready to cut up the fabric.*

• • •

A CHANCE FOR REPRISALS

By Marguerite Martyn

April 12, 1918. "Over there" the French girl may have her innings, but over here foreign officers on various missions have found the goddess at whose shrine to worship.

This all-American "goddess" is a society woman surrounded by good-looking foreigners (who are a bit too dapper, if you ask me). One of them holds a French-English dictionary.

In the bubble at top left, three American soldiers sit at a "Cafe de Paris" with a jeune femme who seems quite taken with two other Yanks passing by. Maybe they are those jaunty athletes from General Wood's 89th Division. Notice the truculent swing of their walk.

• • •

SANTA CLAUS AS CENSOR

Drawn for the Post-Dispatch by Marguerite Martyn.

November 20, 1917. "Her" *on the left is packing a Christmas box for* "Him, somewhere in France." *Santa Claus has already checked the list (twice) and has vetoed an umbrella, walking stick, and foot slippers.*

Still left to get rid of: Copy of Shakespeare, gold pen knife, photo in frame, smoking jacket, collar box, reading lamp, gold pencil, gold key ring, cravats, scarf pin, belt buckle, ash tray, *and* sofa cushion. *Santa might approve of the beribboned package at bottom left, marked* "To Sammy from Susan," *depending on what it contains.*

He will definitely allow the "Photo in frame."

26. The Snipers

THE RELIEF GOES OUT

By Clair Kenamore

May 6, 1918. *(Published June 30.)* The regiment had held a position for six days, and it was time to give relief to our men in the trenches. Reliefs are made at night. The men say it always seems to rain on those nights, but since it is always raining in this part of France in the spring, it doesn't matter.

Each man carried his full pack, all his worldly goods were with him. His jaunty little forage cap was stored away, and he wore his steel helmet, for all these roads were covered by enemy guns. The German held this country for a while, and his maps of it are as good and accurate as ours.

Through the sodden night came the occasional booming of guns. Sullen and dogged they sounded, but not terrifying. They were miles away, and their intended targets might be any place in a stretch of a dozen miles. The men feel uneasy when our guns do not answer, but loosen up when they hear a return, shell for shell.

Slogging in the muck on a dark night is not as bad as it sounds. Your shrapnel helmet is an excellent rain hat. Your overcoat is warm and will turn water for a long time, and over that is the canvas half of a tent, which makes a bully raincoat. Any soldier may have shortened the skirts of his coat about a foot, in violation of orders. He can march a lot better if that extra length is left behind and not flapping about wetly over his calves.

Besides pack, haversack, and rifle hung over his shoulders that night was his emergency ration, something he disdained in camp, but now very sweet in prospect, into which he expected to forage at the first halt.

But the powers that made the rules had other ideas, and they decreed that the topmost item should be the American gas mask, supplemented by a French mask, which was light and convenient and carried for emergencies.

Warning Is Passed Down the Line

A sentry, dripping with rain, appeared from the dark and halted the marchers. He talked with the captain. A warning was passed down the line: "Gas alert!"

From this time until we return to this point on our journey a week hence, the gas mask must always be carried at the ready. Waking, sleeping, eating, washing, shaving, afoot or on horseback, in lightning, thunder, sleet, and rain, officers and men wear the canvas bag hooked up well on the chest, almost under the chin. From there it is a matter of seconds to pull it over the face and breathe.

There were some muffled profanities, but everyone did as he was told.

This change was a nuisance. Legs had become accustomed to the marching pace, and if one can keep the upper part of the body fairly still, the discomfort was not great. But with the swinging of things about to pull up the mask and hooking the leather tab over the brass knob, unexpected and unknown damp places were discovered. Certain unacknowledged wearinesses became apparent. A trickle of water ran from the edge of the helmet down the neck.

We passed through a village where three posts halted our leader and demanded the countersign. Then the liaison officer from the front guided us along a dreary, camouflaged road that skirted a soaked field into a second town where all whispers were hushed and the companies were separated.

Through a Barn and Into a Trench

Ours went into a barn-like stone building, out at another door, and there we were — in a trench.

In the darkness and rain it appeared far from a pleasant place, and it seemed deserted. The first man we saw was lying on the parapet, keeping the ceaseless watch on No Man's Land. He was the color of everything else, the shapeless, blobby gray of a rainy night.

The line of men waiting to be relieved flattened themselves against the side of the narrow trench to let their successors pass. In the jams and halts, soldiers of the incoming and outgoing units talked in whispers.

In story books the departing men would have uttered words of cheer and encouragement. But that is not the nature of a soldier. Does he say, "Good luck!" or "Carry on!" or "Forward to victory under the hand of God"?

He does not.

One whispered, "I'm sorry for you poor fish. The Boche comes in here whenever he wants to, and he's got a new gas that you can't smell 'til it knocks you over. Be awful careful, 'cause this is the worst piece of trench on the whole line."

Our incoming man, with wet feet and a trickle of water down his spine, with sixty pounds on his back, standing on a crazy, slanting, soaking length of duck board, stood and listened.

He replied:

"You better get back and warm up, bub. There won't be no Boche take it while this outfit is here. This is a bunch of he-men coming in."

New Defenders Shown Their Places

The squads pushed on to their appointed places, guided by their liaisons. A man from each squad remained with the newcomers, to show them where everything was stored and to lend a helping hand.

The sentries were relieved. They are the silent figures on top of the parapet, in the concealed advanced stations, or in the listening posts.

At midnight, supper was brought up. The retiring outfit had left their cooks and bearers to perform this hospitality and to show the new men how it was done. Coffee and stew, both steaming hot, and plenty of "seconds."

The stew was of beef, potatoes, and other vegetables. As a satisfying, nourishing, and palatable food, it was a work of art. The white bread was excellent, and the coffee was very good, as Army coffee always is.

But that thick, rich, stew, with an occasional piece of bacon floating about as a surprise, was something to stick in the memory, to be recalled in whispers at the reunion banquet at home, in the dim future.

Supper was over, and the sentries were fed and replaced. The food bearers, with their marmite (heat-retaining) cans, had retired. I was in a dugout with six other men, knees under chin, hunkered on a wet bunk. Two men might have been comfortable there, but no more.

We inhaled the mixed odor of an abandoned cellar and decaying leather. We slept.

At 6 p.m. seven men gathered in the captain's billet, a pleasant old house which was also company headquarters.

These were new men to us. The company on the front line does not do the raiding and patrolling. Support teams do that.

Try to Stay Awake

Holding the front line is not always the worst of work. Often it is the best. The companies behind have to bring up your supplies. Your business is to stay awake for six days and nights, most of the time, and to stay wet, sleep with your clothes on, and fight Germans and lice — if any appear. The lice will do so without doubt.

The patrol party had all its instructions when it arrived. It had as complete a knowledge of the ground as could be obtained by maps and a somewhat distant observation by day.

They were to pass outside our wire, check the little graveyard thoroughly and make sure there was no German sniper hidden there. They were to leave our own sniper, then come back home.

The man we left was a sergeant of twenty-eight, a school teacher before joining up. He had a sharpshooter's medal. He was not keen on this assignment. He was to lie motionless night and day, unless he caught sight of an enemy, then he would down him.

That is, if he got the breaks and the other fellow did not see him first.

Our shrapnel helmets are painted with what is presumed to be a dull, unglazed paint, but they still reflect the light, especially where the color has been knocked off. So we covered them with cloth.

Each man removed every possible mark of identification. No neck tags. Pockets emptied. Collar markings gone. If any was killed, the Boche would find nothing on the body to identify his enemy's units.

Each man blackened his face with burnt cork. Luminous wrist watches were pushed well up the arm and sleeves pulled down to cover them.

Efficient New Trench Knife

Each man carried a rifle and 120 rounds of ammunition. Each had a wonderfully efficient new trench knife, a most malicious weapon worn in a belt scabbard. It is fifteen inches long, quite enough to reach the innermost consciousness of any German. It has three sides, like the old bayonet used in the Civil War. It has a grip with corrugations for each finger and a steel guard over it with knobs like a pair of knuckles. If you miss him, when you gig at him, you can still bring him down when the guard gets him.

Seven men and a lieutenant set forth, all garbed alike. Officers wear no distinguishing marks in the front line, and they carry rifles, not pistols. The enemy likes to snipe officers *(who were normally armed with pistols).*

Down the trench which cut across the once-pretty lawn of the house the captain had taken over. Both ends were protected by out-buildings, but German machine guns had plowed up the flower beds and smashed the statue of a naked boy holding a fish.

By 9 o'clock the first man, the "ground scout," had passed through a hole in a protecting hedge and snaked his way toward our barbed wire. The plan was called the "accordion" movement. The first man advanced twenty-five yards and waited. The second man crawled up. Then the third, the lieutenant. The sniper came fourth, then the other three. When the accordion was all together, it was drawn out again, and the party moved in twenty-five-yard increments.

It is hard work, pushing a rifle in front of you, the muzzle filled with a little roll of cloth to keep out the mud. Belt buckles catch on weeds.

The stretching and contracting column reached the graveyard, then each man was on his own until the other side. The lieutenant found a place for our sniper, with last year's and this year's weeds in front. The ex-teacher was left there in quiet, to think his long thoughts and wait for the dawn.

• • •

As the patrol walked back through the silent town, a faint moon tried to break through the heavy, scattering clouds. A thin sort of feeling pervades between 1 and 2 in the morning; physical weariness adds to it, and rain helps it along.

Shell holes every hundred feet or so told us how well the enemy guns had found this road to be an easy target, and the fragile fabric of the camouflage stretched alongside was no protection from an enemy artilleryman firing from a map and according to a geometrical rule of gunnery.

There was comfort in a long line of French artillery drawn along the road near the town where the supporting troops were quartered, and joy in the hot coffee the patrollers drank while removing the cork from their faces. Then the tired soldiers slept, wet but content.

A CHILLING DISCOVERY

May 7, 1918. *(Published June 30.)* It was a day of steady spring rain, and at 8 p.m. the same lieutenant led seven fresh men on patrol to pick up his sniper from the night before and repair some wires where shells had broken them.

They crawled carefully to the graveyard, accordioned their way across it and called out cautiously in the agreed signal. No answer.

He must have fallen asleep, the lieutenant thought. He crept up, found the man, and shook him. But he was wasting effort. The teacher was dead.

The lieutenant sent the body back and proceeded to fix the wires as planned.

Back in the dugout it was found that a bullet had caught the American in the top of the head and passed lengthwise through him. A German sniper had got his blow in first.

27. March Into the Mountains

THE TROOPS OF THE 35TH DIVISION moved to their assigned sector on the Western Front, "somewhere in France."

MIDWEST TROOPS HEAD FOR BATTLE

By Clair Kenamore

June 17, 1918. *(Published August 11.)* From the Middle West they came, from Missouri and Kansas. The twelve hundred American troops were brought sixty miles in motor trucks from the little French town where they had rested for four days. And at 6 p.m. that day we passed into what was Germany, now reoccupied by France.

Today marked one of those gay occasions which tempers the bitterness of war. It was payday, the first in three months.

Madelon is not the name of the town, and for the present its true identity must be hidden. It lies in the Alsatian hills which France lost in 1871 and regained three years ago.

The war is over, so now we can say. It was named Wesserling. Rich with coal. (Postcard photo, next.)

8 WESSERLING   Vue sur Wesserling

• • •

In this pleasant summer land of Alsace, there are many girls with yellow hair whose mission in life is to stand by the roadside and wave at soldiers. In return, they heard such hearty Midwestern greetings as "Oh, you kid!" and "Hello, Red!" and "You just wait till I come back, Blondie!"

When the chubby, red-cheeked girls cried *"Américains! Américains!"* and threw flowers in the trucks, soldiers answered, "Me for you, baby!" And so they made friends with the civilian population, as recommended in the field-service regulations.

• • •

Officers were billeted in private homes and the men slept in a large barracks which had once belonged to the Kaiser's army.

June 18, 1918. *(Published August 11.)* The next day our men found a pretty town nestled between two mountain chains. Its chief business in peacetime was to be a playground and summer resort. There were many little hotels, cafés, and souvenir shops, and in these places lay a new and charming discovery.

These soldiers, formerly of the Missouri National Guard, were largely of families bearing German names. Possibly a third spoke German. They were, of course, all good Americans, but they had no French, and their assays with that language during the past three weeks had been most trying.

But here, in this happy valley, everybody spoke German, and the *cordiale* was immediately established. The strapping, khaki-clad Americans and the blonde shopgirls in white aprons lifted up their voices and denounced the Kaiser and all his works. In German.

The Alsatians marveled at the height and width of the Americans, and the latter told them to "Just wait until we get into the trenches, and you will see something. Just you wait, that's all."

Our soldiers bought postcards and chocolates and went happy to bed.

In beautiful sunshine, the fresh air blew from the pine trees, and all around rose the mountains in which we were to meet the foe. Every man was told to sleep as much as he could, to prepare for a hard night march.

So far as I know, nobody slept. There were shops to visit and pretty blondes to talk with.

June 18 and 19, 1918. *(Published August 11.)* At 8 p.m., inspection — packs, helmets, gas masks, rifles and everything else that goes to make up the sixty-five pounds of a soldier's burden. Every man had more than that, what with chocolate, little round cheeses, and souvenirs.

At 9 p.m. the column set off for the high hills, swinging like tried fighting men along the street. Townspeople and French soldiers lined the *(Wesserling)* roadside to cheer.

For nine miles it was a perfect road, with a gentle down grade. An easy march of twenty-five minutes. Five minutes of rest. Repeat.

Then came the foot of the mountain, where the division first showed its strength. The climb began. The men will never forget that awful march, into the mountains and down the other side. The swing to the foothills was accompanied by singing, jests, and laughter. Platoons marched with intervals of twenty paces between them. *(Next photo, by Kenamore.)*

None of the new war songs were particularly good marching tunes, although all were tried that night. The most satisfactory were the old, unprintable Army songs, like "Down in the Valley," "Heigh, Ho! The Rolling River," and such.

The climb began. The road is cut in the side of a mountain, and it zig-zags back and forth, like the switchbacks in the Rocky Mountains. It was a terrific grade for troops carrying full pack, but the men settled themselves more snugly into their harnesses and pushed ahead.

Some inextinguishable spirits kept up the songs, but by 11 p.m. the order came that songs and cigarettes would be prohibited.

As the road crept up from the valley, the long column, still marching four abreast, got into regions where the night winds played. The sky was overcast, with scudding, broken clouds, through which the westing moon sometimes broke for a second, to be covered again.

Halts were more frequent: fifteen minutes' march, with five minutes' rest, for the grade was steep and the step slower.

There came a point when human strength exhausted itself, and it coincided with a whistle blast from the head of the column which meant "Fall out!"

Legs became weary and packs assumed prodigious weights, and the gay lads who had philandered the sunny afternoon away in the town may have been regretting the sleep they did not get.

Lights Twinkle in the Valley

Just at midnight, the moon broke through the clouds and the road took a level stretch. Tired muscles rejoiced, and glad eyes looked upon the majestic mountains behind them. The valley was spread out like a tiny picture, and lights twinkled in the toy towns above which rose the great mountains of the Hautes Vosges. White dots were chalets and farmhouses.

Ahead, blasts of mist whipped across the trail — refreshing if the column was toiling on, but chilling if it blew while the men were at halt. The mists were puzzling for a time until it became clear that they were the fleecy clouds so much admired the day before from the sidewalks and cafés down below.

I was struck by the realization that these Middle Westerners from the lowlands of the great Mississippi had mounted into the skies.

About 1:30 a weird and somber forest swallowed the winding trail.

Few of the men had ever seen such a forest, where erect fir trees, with boughs hanging downward, rose more than a hundred feet and formed a ceiling so dense that all vegetation on the ground had died for lack of sun. It was like a great, dim cathedral at midnight, and the roadway like a narrow, hopeless thread.

Through this eerie wood they labored, still on the killing upgrade. Heavy rain fell. and wet, cold clouds blew through the ghostly pines. When the column stopped, the men threw themselves on the ground, or they sat on the uphill side of the road, their backs against something solid. On the downhill side there was usually a parapet, but beyond that was a sheer drop to unknown depths, and it was no place for a sleepy soldier to sit.

At two-thirty it seemed that the limit of human endurance was reached, but from then on there were deeds worthy of medals.

Captains and lieutenants were marching with their men, but it was the enlisted men who looked out for each other.

Carries Another Man's Pack

Many a private carried another man's rifle or gas mask or helmet to relieve him. One sergeant lugged his sick "bunkie's" pack as well as his own all the way. That means he carried more than a hundred pounds up that terrible mountain.

Some day, I vow, I will give his name, which is so German that it is hard to pronounce, but he is as American as biscuits for breakfast.

He talked to every man in his platoon that night, and his counsel was never the same for any two men. Some he ridiculed, some he cursed, some he encouraged, some he abused, some he laughed at, some he sympathized with, but he brought every man through to the end.

(At inspection the next day, each soldier had his equipment, even though some of them might have lightened their packs by "losing" some of the contents along the way. This sergeant had hung on to enough energy to browse around and "griffle" a needed item from a sleeping member of another company.)

Well after two o'clock, depression came with the black night and the unceasing toil. A cold, wet cloud blew over the sinuous line. We were all halted at an open place where at a time past somebody had decided he could build a farm and graze his cattle.

A halt was called, and the men threw themselves prostrate on the road without loosening their packs. At that moment the outfit badly needed a "pick-me-up," and it came.

"Listen!" said the sergeant.

Through the cloud and the mournful wind we heard the thunder of our guns — the French 75s. They were talking to us.

In most places, Missouri and Kansas are pretty flat, the Ozarks excepted. Men of the 35th Division were not used to mountains like those in the Department of the Vosges in western France; they could look across the valley below, and in the distance was the enemy.

FRENCH ADVISE TO TAKE IT SLOW

June 21, 1918. *(Published August 11.)* The big guns were booming, slowly and monotonously. From our eyrie, we could see into distant Alsatian valleys, far behind the Boche lines.

French soldiers do not carry their packs up the mountain that the Americans had just conquered. Supplies are brought by an aerial tramway on steel cables. The French looked with amazement at the Americans who had made the march in seven hours with their packs.

Our men fraternized with the French in a wonderful manner. A few of them spoke French, and an occasional French soldier spoke English, and in several cases both spoke German.

Nevertheless they got along fine. They compared rifles and canteens, and the French explained the uses of the spiked walking stick. (The Americans quickly made such sticks for themselves.)

They felt each other's muscles and said, "Très bon!"

• • •

At 9 p.m. two companies were lined up for another hike, the short four miles to the trenches. Two companies were left behind, envious and snarling and highly critical of the officers who made the decision on who would fight and who would not.

At 10 o'clock the favored companies moved toward the front. From the valley the roar of sullen guns rolled up like billows, so there was no need for silence, and the men were allowed to sing the old, reprehensible songs until the firing ceased and the order of silence was passed down the line by runners.

On the march up the mountain, intervals of twenty yards had divided the platoons. On this night, the intervals were lengthened to two hundred yards, and each platoon had a French guide.

The brigade went into the trenches. At some places a squad of French would be joined by a squad of Americans. In other places, there would be four men from each army.

In many instances the men from both armies would not have six words in common, and it was pitch black, and nothing must be said above a whisper, for the Hun was less than a grenade throw away, yet they amalgamated like friends and brothers in arms.

On every sentry post an American joined a Frenchman. They both looked into No Man's Land and the dark mass of barbed wire.

Tri-Lingual Conference

Back at a row of huts along the mountain side, the officers conferred, partly in English, partly through interpreters, and partly in German.

A young lieutenant, fresh from a year of hard training in the States and full of vigor and ambition, allowed that it was too late to do any action this day, so it might be well to wait until tomorrow night before beginning trench raids, patrols, and all that.

The American major listened patiently and explained to everybody that the French were in charge and they showed no desire to "start anything." If they left Fritz alone, he would leave them alone, and that was a very desirable thing.

So their French hosts did not make a conspicuously favorable impression on the half dozen young officers who slept in a little billet cut in the mountainside that night.

"No wonder these French have not won the war," one lieutenant said. "They don't want to fight. Wait a while; we'll show them how to whip the Boche!"

Next day at dinner, there was a change of view.

'No War Will Be Decided Here'

The French captain at the head of the table said:

"My battalion came into this sector just four days before you. We came to rest. We came to heal our wounds and teach your men how to fight.

"We have been in the Chemin des Dames *(a chain of heavily contested wooded rocky ridges),* where we were in a hard battle *(the Third Battle of the Aisne).* Our regiment had many casualties, and we are very tired.

"We caught the full brunt of the first rush of the Hun. The major in whose place I sit was killed. Our colonel was killed or captured, we do not know which.

"The flag of our regiment has been decorated for gallantry by our commanding general. We desire to refill our ranks, regain our strength and go back to the battle line, where the war will be won or lost.

"We do not want to waste our strength in these mountains; no war will be decided here."

He looked around him.

"My two companies in the line with your men are commanded by lieutenants, as both captains were killed. The lieutenant who is sitting there beside you was a sergeant yesterday."

The American major looked upon the shamed faces of his lieutenants and replied:

"Your orders shall be our law. We come here to learn from you. You have paid dearly for your master's degrees."

Silent Agreement

The French captain was a handsome man of 35, whose tunic was more or less unbuttoned and whose shirt had a constant inclination to crawl from under his belt. He was forever dashing off to the top of the mountain, or sliding down into the valley, to see how things went in the trenches.

At dinner the next night he displayed the medals which he kept in a pocketbook.

He had the Croix de Guerre, with three palms and four stars, indicating three citations from his army, two from his division, and two from his corps, testifying to a career of personal bravery which few have achieved.

He also had the Legion of Honor.

Afterward, there seemed to be silent agreement to reserve further criticism of the French until the critics had at least a few citations to match these.

• • •

Along the Chemin des Dames in the Aisne today you can find cemeteries for French, German, British, Danish, Italian, and American fighters. The Danes, from Germany's Schleswig-Holstein province, had fought in the German army.

Next: How the folks back home sustained these troops.

28. How to Support the War Effort

IN ST. LOUIS, THE RED CROSS was intent on raising funds. It orga-
nized a day for women volunteers to go downtown and ask everybody
to ante up for the Good Cause.

At least one passer-by suddenly became interested in volunteering
(next drawing).

The idea was to inveigle the men into contributing, because they were the ones who controlled the wealth.

SILENCE ABANDONED IN SEARCH FOR FUNDS

By Marguerite Martyn

May 26, 1918. Whoo-ee!

What a lot of good-natured sarcasm did I hear from the men on the streets yesterday. But none of us women cared how much they fussed, just so they dug deep.

The men complained to one another.

"How much did it cost you to walk down Olive Street?"

"It'll be cheaper to take a cab back."

"Which is the safest way over to Washington Avenue? Via the tunnel, I guess."

"They ought to send these women over to get the Kaiser. They'd annoy him to death."

St. Louis has known street collections, but nothing like this one. It was to have been a "mute appeal." Orders were that no word was to be spoken, only the presentation of an official Red Cross money box.

Oh, it was almost pitiful early in the day to see women "talking" with their eyes and gestures, while forcibly restraining themselves from actual speech.

Quickly enough, everybody agreed that it could just not be done, and soon the women were not only asking out loud but also employing whatever other gifts were at hand to attract attention.

Mrs. Claude Matthews, at Twelfth and Olive, revived a childhood skill at baseball to catch coins tossed from passing streetcars and automobiles. And when the crowd discovered that here was a woman who really could shag a nickel or a dime, they kept her busy. One man drove around the block several times, tossing her a coin on each pass, and she never flubbed it. *(Next drawing.)*

MRS. CLAUDE MATTHEWS
DISTINGUISHED HERSELF AS A CATCHER

Contributions came so fast that it was impossible to frame a different line for each person, so at some corners you heard "Give again!" or "Give 'til it hurts!" over and over. I did not know how many varying inflections and what a world of dramatic appeal could be put into such simple phrases.

Mary Hendry reaped great rewards. She said she was successful because she waited outside a popular-priced restaurant until the "brute" was fed and in good humor before she approached him *(next image)*.

"BUT YOU HAVEN'T PUT ANYTHING IN MY BOX—"

MISS MARY HENDRY
STATIONED HERSELF OUTSIDE
A POPULAR RESTAURANT AND
CAUGHT THEM AS THEY CAME OUT
"WHILE THEY WERE IN A GOOD HUMOR."

But another woman did just the opposite. She took her position before a restaurant and appealed to the men going inside to divide with her "50-50" what they expected to spend for their lunch. And this was an expensive place to eat, too.

Cashiers in all the stores and banks were soon bereft of nickels. A popular strategy for a man to get past the persistent appeal was to change a large coin into pennies and poke one into each box thrust at him. Some were perfectly frank about it and walked down the street counting aloud, "eighty-five, eighty-six," etc., as they went.

There was much humor, but there was some pathos as well, for the donation of a coin or two was a serious matter for many a contributor. For the most part, though, there arose a spirit of gayness and human friendliness.

Anybody who passed Miss Lillie Hannegan at the corner of Eleventh and Olive could not evade the experience if she merely suspected that he hadn't contributed. She is the sister of our new chief of detectives, and not a guilty man escaped until he gave what he could. *(Next sketch.)*

Judge Selden Spencer in an imposing uniform drove by in an imposing car and distributed lunch money to the soldiers on duty by throwing a 50-cent coin to each man instead of stopping and making a ceremony of it.

"I never saw such heartlessness," said a woman at Sixth and Olive, "throwing money at these men as if they were tramps or beggars."

"But you don't seem to object to having money thrown at you today," I reminded them. "Oh, no!" they didn't take issue with that.

• • •

WONDERFUL EXPERIENCE AMONG THE POOR

May 28, 1918. Below is an individual experience of a Liberty Bond saleswoman, who sent it to me in the hope it might inspire other women to dig into their purses and double their subscriptions in the current drive.

Her impressions echo what is becoming a complaint from headquarters — that rich women do not show up as well in the returns as do working women or wives of salaried men.

The letter-writer had gone out with a friend to what were called the "congested districts."

"Rose and I were doing a poor precinct, a large one (from X Street to the wharf), and we approached our task rather hopelessly, but I want to tell you that what we did not gain in money we did gain in confidence.

"We started out when the whistles blew Saturday *(to mark the end of factory work on that day),* and we hadn't canvassed but two houses when we had lost all our anxiety. We had sold two bonds in half a block. At the other homes, the women said their husbands were taking them at the job and they couldn't afford any more.

"Now, that might not seem much to you saleswomen who have better precincts to work, but if you were to sell as many bonds in

WE SELL
$10
LIBERTY LOAN
Certificates
Five of These Buy a $50 Bond
AN EASY WAY TO SAVE
FOR A LIBERTY BOND
Subscriptions Taken Here
U. S. Government
$3\frac{10}{2}\%$
LIBERTY BONDS
Do YOUR Part to
End the War!

comparison to the wealth of your precinct as the poor have bought in ours, I am sure the city would easily double its quota.

Brothers Away In Service

"We went into one back yard and found a girl of about 20 scrubbing the steps. We started to tell her that we represented the Liberty Bond organization, but she interrupted —

"'You know, I've been thinking and thinking about whether I can take a bond, and I just don't know how it's possible. You see, all I do is to keep house for my three brothers. Two of them are in France, and the third is waiting for his call. I'll be alone then, and of course it will then be easy for me to get work in these times.

"'Could you come back Monday? It almost breaks my heart to say that I can't take a bond.'

"'You think about it,' I said, 'and I'll be back Monday.'

"Then I thought of all my acquaintances who believe they are doing so much when they buy a fifty-dollar bond, while this poor girl does not feel she is doing enough. Oh, doesn't it just make you fighting mad?!

"Noticing my service pin with its two stars *(like the one in the preceding image),* the girl said, 'Have you someone gone, too?'

"I told her about my brothers, and she replied, 'Wait a minute, I want to show you *my* brothers' pictures!' She brought them out, and Rose and I admired them.

"'Lordy!' she said. 'The boys are doing so much, and we can do so little. Give me one of those coupon books quick. I'll be getting a job, and I'll pay for it somehow.'

"Oh, how glad I am that my own brothers are fighting for a woman like her!

Will 'Pay for It Somehow'

"Now, I had decided that I couldn't take a bond at this time. I still haven't paid off my third loan, and I have pledged for War Stamps. And I thought I just have to have new shoes and a hat. But how would you feel if you had just sold a bond to a woman like that?

"Well, I have an old pair of shoes I can have patched, and my last year's hat will do, and I'm going to buy a bond and 'pay for it somehow,' too. And I hope the dear Lord and my brothers in France will forgive me for hesitating until now.

"I could tell many more stories about how the poor people are buying bonds — about the boy we visited who was honorably discharged from the 138th Regiment because he was crippled while on guard duty in this city and who is turning his insurance into Liberty Bonds.

"Why, I wish you could see the little old room he is living in! We shook hands with him, and Rose said afterwards, 'I wanted to kiss the stumps where his fingers were and say, "The women of America thank you!"'

"I just wish some of you people who are sitting in your nice, comfy homes could sell bonds in our precinct and see how many the poor are buying!

"And there is hardly a foreigner living with his family in the dingiest little home who hasn't bought a Liberty Bond to help his adopted country.

"There are many of us who have experienced all the blessings of this dear land who are not buying as these foreign-born people are. I mean comparatively, of course.

"They are a sermon to us, if we love this free, beautiful country. We should buy bonds to keep it that way for our children, as our forefathers kept it for us."

THE ARMY AT HOME

April 16, 1918. Anybody may enlist under this Flag. There are no age limits, no physical tests, no discrimination of sex, color, race, or creed. Anybody may join the Liberty Loan procession. *(Opposite.)*

29. America's Battle Failure

THE 35TH DIVISION MOVED INTO the Meuse-Argonne front in mid-September 1918, preparing for battle. For reasons nobody knows (but can guess at), on September 21, just five days before the offensive began, Major General Peter Traub took the opportunity to fire two brigade commanders and four regimental commanders, all of them from the National Guard. According to military historian Edward G. Lengel, the replacements had no combat experience and did not know the men in their troops, but they <u>were</u> Regular Army.

One of their tasks: Take a heavily fortified Vauquois Hill, which the French had already tried to capture three times and were calling "The Hill of the Dead."

It was the beginning of the Meuse-Argonne attack, "America's Greatest Battle," involving 225,000 untested American soldiers. The 35th was placed on the right bank of the Aire River, and the 28th, from Pennsylvania, was on its left.

TOWARD THE HINDENBURG LINE

By Clair Kenamore

September 28, 1918. *(Published September 30.)* The enemy has brought strong reserves from his depleted divisions to oppose the eastward progress of the Kansas and Missouri troops down the River Aire.

After a hard day's fighting yesterday, the men pressed on all night, and by 11:45 this morning they had reached the objectives set for them. They might have gone further, but the difficult terrain and stiff resistance delayed their advance to such an extent that they found themselves in a pronounced salient, far ahead of supporting troops on each side.

Through the night, they went with rifle, pistol, bayonet, and grenades. In daylight, though, most of the enemy machine guns were taken out by our tanks.

On our left, across the river, another American division fighting through the Argonne Forest was held back, and the German machine guns and artillery in front of them were thus able to shoot up our troops on this side.

Our men then crossed over to aid in reducing a town where defenders fought bitterly. They struggled straight ahead, through shell-fire, over barbed wire and concrete dugouts of tremendous strength, past machine-gun nests and, at dusk, they were digging themselves in to withstand a night of barbarism, raids, and counterattacks.

This morning our troops attacked again at 6:30 in cold rain and found the enemy everywhere, but the spirit and dash of our men carried them forward.

The opposing troops are of the Prussian Guard, the choicest of the Kaiser's army, but the Americans went ahead through the mass of entanglements and defenses known as the Hindenburg Line, and it was breakfast time, and there was cold rain falling. But details like that are for history. Nightfall found them at the appointed line for the end of the second day.

Weird, Chilling Mists

Today's fighting was harder, but not so spectacular as on the first day. Yesterday's chilling mists have given a touch of the weird and mystic seldom seen on a battlefield.

The advance at 5:30 this morning was cautious and well-ordered. One village was passed on both sides and left for the moppers-up. Fog grew denser further on.

The enemy in the front positions was shaken by the artillery and made no stubborn resistance, except in occasional machine-gun nests where they fought to the death. These patches were reduced by tanks or squads creeping upon them with grenades and pistols.

The rapidity of the advance and the density of the fog caused some of the officers and platoons to become separated and in some cases to lose direction. Some platoons moved faster than others, and the slower ones found Americans in front they did not know were there.

So dense was the mist that it was like fighting in a thick smoke. Forms would come out of the gloom like figures on a developing photographic plate, and sometimes one could not tell if he was facing friend or foe.

Vauquois Hill and its destroyed town were bypassed and left for the artillery to reduce.

Americans Are Surrounded

A detachment of the Missouri and Kansas men were plunging ahead on a choppy road when a sharp volley from machine guns opened from their rear. The men were ordered to take cover, and at that moment, the fog lifted and our advanced unit discovered itself surrounded. The shrewd German tactic had been to allow the Americans to sift through between silent machine-gun nests. When they were well within the trap, gunfire erupted from all sides.

This was the same textbook tactic the Germans used elsewhere in the same battle, resulting in the storied "Lost Battalion" of Texas troops, surrounded by the enemy for a week with no food — and water available only by crawling to a nearby stream under enemy fire. They were featured in song, verse, and motion picture. The Kansan-Missourians fought their way out of a similar trap and were celebrated nowhere.

The fog lifted, and all was clear, open fighting. The beleaguered troops took care of the ring of enemy machine guns in the long grass to their rear. Most were silenced by rifles; others fell to pistols and hand grenades.

About the middle of this scrap, when the outcome was still uncertain, along the Cheppy road came the tanks at a merry gait. They made short work of the tangle of trees, bush, and wire where a nest had been. Then off toward the town of Cheppy itself, but hardly were they away before a new bunch of machine guns opened up against our troops along the little stream at the edge of town.

The tanks turned sharper than automobiles could and came back "hell for leather." The creek was soon as clean as a bone. No prisoners were taken.

Our men *(the 138th Infantry)* then moved against the heavily fortified town. The fight was our rifles against their machine guns, but by noon it was in our hands, and prisoners were going back.

• • •

The fighting in and for these positions produced many strange results. One officer found his platoon advancing in skirmish line but facing in the wrong direction.

Sammy Goldberg, Colonel Harry S. Howland's striker *(orderly),* became separated from his outfit and hid in a big shell hole to escape from a machine gun in the next hole over. A hand grenade tossed into his refuge failed to explode. He knew that another would follow and probably kill him, so he decided to get a few of the enemy first. When he crawled over, a pistol in his hand, the Germans threw up their hands, and he took seventeen prisoners to the rear.

ST. LOUISAN WHO TOOK 17 GERMAN PRISONERS

SERGT. SAMMY GOLDBERG.

Goldberg, the son of a St. Louis baker, won the Distinguished Service Cross for this feat. (P-D photo printed September 30, 1918.)

• • •

The fog stopped all communication. Lamps and flags *(for signaling)* were useless, pigeons unable to fly, and wires were run slowly and with great difficulty.

Lieutenant Venable of Mexico, Missouri, running a wire with five men, struck a machine-gun nest and lost all of his comrades. Another squad was shot up but finally made it to brigade headquarters, which was in a shell hole. The radio set up in an adjoining hole could not connect with the rear, but heard communiqués passing through and got a time report from the Eiffel Tower.

In such mist, the aviation liaison was almost useless, but occasionally one pilot swooped so low in the rear of the enemy lines that it seemed he would hit a tree or the earth. Knowing that the flier is trying so hard is a great cheer to the infantry.

It was even more cheering when the rain quit after noon today, and the air was filled with fighting, photographing, bombing, and observing planes. Our balloons were up, and whenever the enemy appeared, he drew a swarm of American hornets.

The roads are soggy and slick, but all is going forward — guns, trucks, ammunition, ration wagons, water carts, ambulances, and tanks. Dead horses and wrecked motor transports are among the litter of warfare. French tanks, manned by Americans, stop occasionally to pull trucks out of mud or ditch, but in general they keep moving forward toward the Hindenburg line.

• • •

History did not treat this epic battle well, which ended for the Kansan-Missourians when, decimated and pushed back, they were withdrawn overnight. Scholar Robert H. Ferrell of Indiana University summed it up in a 2005 book with the withering title "Collapse at Meuse-Argonne: The Failure of the Missouri-Kansas Division."

He placed major blame on poor training.

Another military historian, Mitchell Yockelson, wrote that division commander Traub had become "a nervous wreck and exhibited bewildering behavior in front of his men . . . . Carrying a swagger stick for no apparent reason, he was seen wandering aimlessly around the battlefield."

The 35th advanced some six miles in three days, but they paid a terrible toll, nearly half dead or in the hospital and the other half overcome with fatigue. In a monumental relief effort on the night of September 29, the seasoned First Division was moved in to relieve them. They found "mangled wheels, harnesses, and swollen horse carcasses where some shell had found its mark." *(Intelligence officer Thomas R. Gowenlock, quoted by Yockelson.)*

Not to mention the remains of Kansas and Missouri soldiers.

• • •

After the war, Kenamore wrote his observations in his book "From Vauquois Hill to Exermont: A History of the Thirty-Fifth Division of the United States Army."

He shared what could not have passed through censorship during the war.

The division looked more like a band of refugees than a military organization, unshaven, dirty, and haggard. Their clothing was soiled and torn, their shoes muddy and worn out. They lay two or three sleeping together, under raincoats. Dysentery had broken out. It was a sorry band for looks, but it had played a great part in a great battle.

Casualties were 1,204 dead and 5,870 wounded. Medals of Honor were awarded to Captain Alexander R. Skinker, 34, of St. Louis and Private Nels Wold, 22, of Minnesota. Both were killed in attacking machine-gun nests.

The 35th spent the rest of the war, six weeks, on a quiet part of the line, in reserve (next chapter).

30. Joyful Armistice Day

AND THEN — IT WAS OVER. Germany was defeated. There was a revolution. Wilhelm abdicated and fled to neutral Holland. An armistice was signed on Monday, November 11, at 6 a.m. by the French clock, effective five hours later. That was 2 a.m. in St. Louis, plenty of time to get the news into all Post-Dispatch editions that day.

CELEBRATION OF JOY

By Marguerite Martyn

December 10, 1918. A letter from Miss Susan Fritsch, dated November 11, to her family at 5475 Cabanne Avenue, tells how the armistice was celebrated in Grenoble, France, where she is working in a soldiers' canteen.

"November 11, 1918.

"Dearest family: This has been the most exciting day I've ever lived through!

"At 11 this morning we received news that the armistice was signed. At that minute, everybody in Grenoble paraded and put out their beautiful flags. It sounded as if the hundred thousand people here, who hadn't cheered for four years, had suddenly found their voices.

"At lunch our hotel manager treated us all to champagne, and we drank to the soldiers in the dining room. Afterward, we went to the crowded streets, people on balconies shouted and sang the Marseillaise. They flew the Allied flags and threw roses and chrysanthemums to the soldiers.

"At 2 o'clock we had a big American parade; hundreds of our soldiers marched in squad formation and stopped at the public buildings to receive cheers.

"We went first to the city hall, then to the university, the French and Italian hospitals; the wounded came out and yelled *'Vive l'Amérique!'* and so forth. They sang their national songs. It was thrilling.

Celebrating in the Street

"At six o'clock it was dark, and the balconies were decorated with bright lanterns and big flare lights. Everyone held his own private celebration in the street.

"After dinner the excitement continued. First a big crowd of picturesque *poilus* (French soldiers) with their enormous loose coats and black 'tams' marched, many waving crutches and some carrying one another or girls on their shoulders. Then the Italians paraded, and Americans were everywhere. Girls embraced them and pinned on flowers.

"The game of the evening was for all the men to join in a circle around any girl they saw, then dance around her, and kiss her. A typical French celebration.

"Skyrockets soared into the sky, and the big gun on the Bastille (*the fortress overlooking the city*) fired every fifteen minutes.

"Next day — they are doing it all over again. Shopkeepers have closed their stores. Everyone is so gay. They embrace us at each very step. All the small children cry out, 'The war is feenished,' all the English they know.

"This afternoon an old lady came into the canteen and, in tears, asked if she might pin a bunch of flowers over President Wilson's picture which we had in the window.

"The celebration will continue all week. The city is wild with joy, but I wish I were in St. Louis tonight so we could celebrate the *fin de la guerre* together!"

SLAUGHTER HAS ENDED

January 24, 1919. Miss Helen Day was a Red Cross worker equipped to be a nurse's aide. She possibly served nearer the actual front than any other St. Louis girl who went over for canteening. Her letter tells how the armistice came to the village of Fromereville in the Argonne region, not far from where our own Missouri men were in action.

"November 11, 1918.

"Dearest mother — The armistice actually signed.

"Were the sky not ablaze with lights of celebration instead of warfare, and the big guns silent for the first time in all these years, I could not possibly believe it!

"Oh, to have lived through this dreadful slaughter. It leaves me all a-quiver. My nerves are taut, still keyed up for the tremendous effort, the daily facing of tragic suffering, and more to come, always more to come!

"I feel very unnatural, like the silence of the atmosphere itself. The ground under us does not tremble from the big guns stationed all about.

Ambulances Keep Coming

"This past week has been the worst, the biggest rushes, the worst wounded, the most terrible suffering. This very day and evening almost reached a climax. As late as 9 this morning, the troops went over the top again, and our ambulances have been on the go. And after 11 o'clock, when our boys rang the bells in the church — the devastated one in which we slept — and the guns and the booming stopped, the ambulances still came in a steady stream — and all evening, too.

"The men lay in the receiving wards for hours before their turns came to go before our many constantly working operating teams. The chocolate kitchen couldn't keep up. We have just finished working all day and all evening without chocolate and cigarettes for these men lying on their litters.

"I can't describe it to you. They are wonderful in their tribulation. It is so terrible that with an armistice coming there should still be this suffering at the end.

"Of course, they had to keep on, and the casualties were great because the Germans retreated so fast that our boys had to fly over hill and dale to follow them, and whenever they were in full view the enemy had a good chance for parting shots. The men who were carried in could scarcely rejoice over the armistice, so terrible was it that many of their comrades were sacrificed just at the end.

"I shall never forget this day and evening. It shall hurt me all my life. One could scarcely be happy in the midst of this gruesome suffering and mental anguish. And all day long, from 11 o'clock, the rejoicers are ringing our little church bell.

"I cried when I heard the bell, but now, tonight, in my tent, I couldn't cry for a million dollars. No more lives to be sacrificed! These days have been far more terrible than anyone can imagine. I am thankful to have been at the front during the last six weeks of the war. Has it really ended? Oh, I can't believe it."

• • •

Kenamore wrote in his book that after the war

Leaves were granted, and officers and men took trips through Paris to the south of France and wandered about through Nice and Monte Carlo, buying postcards and eating restaurant food,

The wounded came back from hospitals, some limping or with scars but all glad to rejoin the old bunch. The officers from the regulars were detached and sent to other tasks, and the original 35th Division officers took up their former posts.

31. Coming Home

SOLDIERS ARE WAITING FOR A SHIP

By Clair Kenamore

February 6, 1919. Commercy, France. The 35th Division has been ordered to LeMans, the station where homebound outfits are assembled and the priority of their departure is determined. Opinion here is that they will reach St. Louis in April.

The division is being concentrated along the railway in Commercy, with orders for readiness to leave tomorrow. The men are in fine condition, and the prospect of going home has them in a high pitch of enthusiasm. The cold billets, the rain, and the mud have been forgotten.

POST-DISPATCH MAN BACK, PRAISES 35TH

By leased wire from the New York bureau of the *Post-Dispatch*

March 1, 1919. Clair Kenamore, field correspondent for the *Post-Dispatch,* arrived in New York City yesterday on the *Rochambeau,* which was crowded with troops.

All their heroic story must be told, he said.

"It was not possible to give all the detailed facts in cables or mail from France under the severe rules of the censorship," he said. "The casualties in the Argonne were about 7,500, but the deaths will not exceed 1,400. It is impossible to get accurate figures.

"The division is about ready to start back home as soon as their turn comes. It was being assembled at LeMans when I left France. That is a clearinghouse for homeward-bound outfits.

"The demand for an investigation made by Representative Philip B. Campbell caused some comment, but the officers are not discussing the matter just now. They prefer to wait until they are back home and freed of their military obligations."

Governor Henry Allen of Kansas had been visiting a Y.M.C.A. unit serving the 35th. He charged that during the week's fighting in the Argonne (Chapter 29), the infantry lacked artillery support, ammunition, food, airplane protection, and adequate care for the wounded — twelve hundred of whom were left on wet ground without shelter for a day and a half.

Colonel Carl Ristine wrote to a House of Representatives investigating committee:

"We were stripped of blankets and had summer underwear and no overcoats . . . men almost froze to death . . . no ambulances for 36 hours, and then only six to nine small ones to haul 6,300 wounded in six days . . . took as long as thirty-six hours to get to the field hospital, and many died on the way." *(Ottawa Herald, Kansas, February 18, 1919.)*

In Kenamore's view, the Midwesterners were heroes. On March 13, 1919, when the 35th was still in Europe, four months after the armistice, he was back in St. Louis and spoke at a packed Central High School auditorium. A Post-Dispatch reporter was there.

APPLAUSE GREETS TALE OF THE THIRTY-FIFTH

March 14, 1919. *(No byline.)* Kenamore's story related wholly to the 35th Division and supplemented his dispatches. Spectators applauded every mention of the awaited return of these men.

He gave an itinerary of the division's movements, which he couldn't do in the "Somewhere in France" days of wartime censorship.

The division arrived in Le Havre, France, on May 1, he said, and headed the next day for Eu, which was pronounced "You" by the men, but which its residents called something like "Ur." It arrived on May 14.

Kenamore told more about the division's early days in France.

"It happened that the British expected an attack on their front, and *(British)* General Douglas Haig borrowed what divisions General Pershing could lend him, nine of them, and the 35th stayed to June 7.

"That was not a very pleasant experience. The men had to use the British rifle, which they never liked, and had to live on British rations, which they liked even less. It was almost impossible to make the Americans drink tea for breakfast.

Through Paris in the Dark

"June 7 the division entrained, day and night, and four days later it arrived at Arches, on the far side of France. Many men went through Paris at night without seeing it.

"There was re-equipment and hard work. On June 30 the advance section left for Wesserling, in Alsace, which was a part of Germany when the war began.

"Headquarters were at Wesserling more than a month. Then the second echelon moved to Ventron *(Vosges)* on July 7, and the first went to Kruth *(Haut-Rhin)* on August 10, both proceeding on August 14 to Gérardmer *(Vosges)*. The headquarters were at various points in the Vosges from July 7 to September 4.

"Under other conditions, this would have been a pleasant place to spend the summer; it had been a resort region for both French and Germans before the war.

"The 138th Regiment was headquartered for three weeks at Black Lake, surrounded by a ring of villas, which became billets. *(These homes or bed-and-breakfasts in Lac Noir in the Haut Rhin were used for officers' lodging or other military purposes.)* We slept in beds, ate good food, and went boating in the evening.

"The men fished in the lake, though no one was ever known to catch a fish."

Kenamore explained.

"The Boche shelled the place, and when a shot fell in the lake, the men would row out, even in the midst of bombardment, to get the fish which were supposed to be brought to the surface.

"But they never found any fish."

He continued his travelogue.

"On September 4, the division entrained for Rosières *(Department of Oise)*. From September 11 to September 18, the three divisions were at Liverdun, Sexey-les-Bois, and Velaine-en-Haye, which the boys called 'Lying-in-the-Hay' *(all in the Department of Meurthe-et-Moselle)*.

"In Rosiéres, one regiment would march as far as ten kilometers, or six-and-a-quarter miles, for no purpose that anyone could see, and another regiment would march in another direction. They knew that fighting was going on somewhere, and there was grumbling because of this drilling to no purpose, as it seemed. Even most of the officers did not know what it was for.

"Then we discovered that the 35th was the reserve for the St. Mihiel offensive *(September 12-15)* and that it was part of its business to be mobile. Its function was to give the high command the privilege of using the troops as it pleased.

Training Would Have Helped

"Perhaps it was unfortunate that the division was not in the St. Mihiel fight, which was an easy operation and would have given training that we needed later in the Argonne.

"The troops moved to Pasavant-en-Argonne *(Marne)* on September 19, and everybody knew that a new battle was coming. The next day, to Autrécourt *(Meuse),* and on September 22 to Grange-le-Comte *(Argonne)* and Camp Perrin. The first echelon moved to Côte-de-Ferimont *(Meurthe-et-Moselle)* on September 27 and the second to Auzeville *(Meuse)* on October 3.

After the battle —

"The division later spent short periods at Thiaucourt, Condé en Barrois, Benoite-Vaux, Sommedieue, Pierrefite and Rosnes, arriving in Lérouville on November 9. It has been there since, unless it has started for a seaport by now. *(These were villages far behind the lines where the shattered 35th recovered and waited for the war to end.)*

• • •

Not until April 19 did the troops of the 35th Division began arriving in New York harbor, aboard the transport Von Steuben. Ten days later, men of St. Louis's own 129th Field Artillery were welcomed back with a parade through downtown, with throngs of spectators, many of whom had been given the day off from work. (Next photo, from the St. Louis Star, April 30, 1919.)

32. Return to Normalcy

SO THE AMERICAN TROOPS CAME HOME. Everybody was yearning for a "Return to normalcy," as all the politicians were putting it. In fact, Republican Senator Warren G. Harding was elected President two years later on that slogan.

For women's reporter Marguerite Martyn, "normalcy" included —

NEGLIGEE GOWNS AT THE ALLIES' SHOP

By Marguerite Martyn

November 23, 1918. Here are some of the boudoir gowns designed by Mrs. Ford Thompson and exhibited by other young women of society acting as models in a fashion parade at the Allies' Shop.

Negligee Boudoir Gowns Designed by Mrs. Ford Thompson and Exhibited by Society Maids and Matrons at the Allies' Shop

The shop, at 610 North Broadway, is being conducted for the benefit of the American Fund for French Women.

From left to right, first is a negligée made of lilac chiffon, with a fichu effect of turquoise blue satin ribbon with narrow Valenciennes lace. The skirt is banded with rows of turquoise chiffon, and the bodice is of lace and chiffon, on a lingerie effect. Worn by Mrs. Knox Taussig.

Second, a white pussy-willow silk slip is banded about the skirt and bodice with fine, pleated, white georgette insertion. Long, floating sleeves are of white georgette, and over the shoulders is a drapery of Chinese blue chiffon.

About the neck is a line of fur, and fastening the belt at one side is a big pink rose. Exhibited by Miss Irwin Hayward.

Number three, a kimono of orange crêpe de Chine as a foundation, overlaid with transparent orchid-tinted georgette. Chinese ornaments of jade rings, metallic braid, and rainbow silk fringe serve to fasten the robe at either side. Miss Emily Isaacs appeared in this robe.

Number four has a foundation of accordion-pleated orchid chiffon, with a long tunic of sea-green georgette. Pale-green transparent stuff is draped over the shoulders, and voluminous lengths of white chiffon float over the arms. A deep-purple rose accents the pastel color scheme. Miss Nancy Bakewell exhibited this model.

Last figure at the right is a warm dressing gown of white elderdown cloth edged about with bands of swan's down. Mrs. Charles Stevenson, recently of Philadelphia, wore this robe.

Women of the 21st Century might find all this is a bit superficial. But if you've been waiting for your sweetheart or your husband to come back alive from any war, of any century, you might understand why buying a negligee boudoir gown could be important.

• • •

For St. Louisans, the war ended with a massive sale of surplus food at the 10th Ward Improvement Association, probably from that canning factory you read about in Chapter 11. (Next image, from the St. Louis Star, August 23, 1919.)

Buying Food From Uncle Sam in Tenth Ward Today

ALL KINDS OF WOMEN LINE UP FOR GROCERIES

August 24, 1919. The line stretched for half a block — men, women, and children ready to buy canned corn, peas, tomatoes, corned beef, and pork-and-beans at Army prices.

There was so much disorganization at first that one woman fainted.

Local women ran over just as they were — in boudoir caps or sun bonnets, in aprons and in house slippers.

Those who came from afar were smartly dressed, some carried thither in handsome equipages. They stepped out of expensive motor cars, one from a Yellow Taxi.

It was a motley assemblage from all classes, inspired by one great enemy, the High Cost of Living. Women discussed among themselves the advisability of buying canned vegetables at 9 cents a can or at $2.16 a case.

Summer furs rubbed democratically against bungalow aprons, for those feet shod in costly oxfords had to get in line just like the women in house slippers.

Crowd-Crashers Castigated

The crowd was good-natured, but when three men started to enter the store without waiting, some people crashed their baskets on the men's heads before they could explain that they had bought earlier and were coming after the goods.

Two women living in the vicinity arrived late and saw a neighbor near the head of the line. They became at once effusively cordial.

"Oh, hello, Mrs. Schultz; that's fine that you saved us a place!" they called out, smiling placatingly at those behind. The latter let out a chorus of "Get out of there!" "Take your turn!" "Get in line!"

They made such a din that a police sergeant ran up and hustled the latecomers back to where they belonged.

When a successful purchaser emerged with more baskets or cases than she could carry, right jealously did she mount guard over them until she could find some way of getting them home. Neighborhood boys and girls possessing toy express wagons did a thriving business at a rate quickly agreed upon, five cents for two blocks' transportation.

But, "To those who hath shall it be given," and it was the richest-looking automobiles that carried away the largest, most enviable purchases.

Like this limousine from a December 15, 1913, drawing (with Martyn offering help from the side):

33. They Never Owned a House

WE CAN ONLY GUESS HOW Marguerite and Clair might have celebrated their reunion when he arrived home from reporting on the front.

The guns of this particular war and the noisy celebrations of its end have faded into history, as have the lives of the two married journalists. All we know of them now is what can be unearthed in some old records and microfilm.

They both had kinship upbringings, close to their parents and siblings. They always lived in a family home or they rented. They never owned a house together.

They started off their married life in 1913 in the big house in Webster Groves with the rest of the Martyn family. (Mother Frances and brothers Phillip and William were living there in 1910.)

While Clair was away in Europe, Marguerite lived in that house on Lake Avenue, across from the sweep of lawn that flanked the Eden Theological Seminary.

When Kenamore came back, they moved around. Once they were at the Y.M.C.A. Hotel on Locust Avenue in St. Louis. Sometimes Clair lived by himself, though he may just have taken a room convenient to the office, so he wouldn't have to travel at night out to the suburbs. Maybe he wanted a retreat to write novels or short stories. We don't know.

In 1927, the Post-Dispatch gave Kenamore the task of soliciting contributions from notable Europeans for the newspaper's Fiftieth

Anniversary edition. He went to Europe with Martyn and his teenage nephew, Bruce.

Marguerite and Bruce returned home, and Clair traveled as far east as Moscow. He came home with articles pledged to be written by Albert Einstein, Bertrand Russell, and Maxim Gorky, among others.

The 230-page 50th Anniversary edition was printed, chockablock with ads, on December 9, 1928, Martyn had a bylined story on how women's fashions had changed over the decades. No sketches.

In April 1930, the census-taker found Kenamore and Martyn back in Tucson. Do you have a radio? No. Are you a veteran? Yes, Clair answered.

The 1931 St. Louis City Directory had them living at that old Locust Avenue address, under its new identification as the Galesworth Hotel. But the next year Kenamore was again in Tucson, for his health. Martyn spent some months at his side in 1932, then returned to Missouri.

• • •

Kenamore became a patient in the Open Air Sanitarium in Milwaukie, Oregon, where he died on November 3, 1935, at age sixty. Dr. Ralph C. Matson, the sanitarium owner, scribbled "chronic pulmonary"-something as the cause of death. Clair was buried in his family's plot in Cedar Grove Cemetery in Dent County, Missouri.

• • •

Martyn stayed at the Post-Dispatch until 1941 — thirty-six years on the job. After eight years in retirement, she died at age sixty-eight on April 17, 1948, in the Webster Groves house. She was laid to rest with her mother and first brother in Oak Hill Cemetery, St. Louis County. Her second brother joined them later.

Appendix
The 92nd Division

WE SAW IN CHAPTER 13 that all draftees of every race and nationality began training at Camp Funston. But to make room at Funston for a separate black regiment, the white men from Missouri and Kansas were packed off to Camp Doniphan to form the 35th Division.

Eventually the Army gathered all the black graduates from the nation's training camps into Funston to form the segregated, and storied, 92nd Division.

Clair Kenamore never wrote about any black Missourians on the front lines because he never saw any. Fortunately for history, the Post-Dispatch's first Washington correspondent, Charles G. Ross, took it upon himself to report on the "Camp Funston negroes," as they were sometimes called.

INVISIBLE COLOR LINE AT TRAINING CAMP

By Charles G. Ross

January 5, 1918. Across the eastern portion of Camp Funston is drawn an invisible line, over which enlisted men may not pass without permission. This is the cantonment's color line, separating the white troops of the 89th Division from the negro troops of the 92nd.

The buildings are exactly of the same type as those occupied by the white men; the streets are as noticeably free from trash; the barracks are quite as neat. Well set-up negroes are seen doing guard duty and working in the headquarters offices. A visitor asking a question of a negro soldier will bring an intelligent and courteous response.

The Army has made the same provision for recreation in huts of the same pattern being maintained by the Y.M.C.A. and the Knights of Columbus. On this side, the recreation is superintended by a negro secretary, and there is well-attended basketball court.

"Many of our boys," the secretary said, "are from city high schools, and they know the game. We have some good teams here."

There's also a football field — the scene, I was told, of many hard-fought games last fall.

There are now twenty-nine hundred negroes here, drawn from all parts of the country. Will they stand the gaff of trench work in France? I asked Major R.P. Harbold, personnel officer.

Negroes Good Soldiers

"Undoubtedly," was the quick response. "I have served with negro troops for thirteen years. I want no better men with which to go into action. Go out and take a look at the men walking about here — lithe, powerful men. How would you like to see a bunch of these men coming at you with bayonets? The Germans are going to find it unpleasant.

"There are ten million negroes in the United States, and we could put between 250,000 and 300,000 of the finest physical types. And the more we put in, the better — for the country and the negro race. They appreciate the opportunity this war has given them. You can see that in the way they conduct themselves here."

He called to a negro officer at an adjoining desk. "Get me the papers on Homer B. Roberts of Kansas City."

Roberts had been an amateur telegraph operator. He wanted to enlist, and he made two or three trips to Camp Funston to find out the procedure. Last December 10 he appeared at the division's headquarters with forty men in tow. All were negro chauffeurs.

Thirty-nine of them passed the physical and were enlisted as ambulance drivers.

"Does the negro readily obey the commands of negro officers, the lieutenants and captains?"

"Yes," Harbold replied. "One little problem came to my attention. It seemed that the negroes in a certain company felt that officers were showing favoritism in selecting non-commissioned officers, almost all of whom were from St. Louis.

"I could have told you the reason without even investigating: The St. Louis negroes got the jobs because they were more alert mentally than the country men. That is generally true of the city negro, with a complex experience of city life behind him. He is simply keener than the country man."

I wondered if any private could have stood out from the ruck, so I asked about Andrew "Boodler" Brown, the star player of the Columbia, Missouri, champion negro team, the Athletics.

"Do I know Brown?" echoed the major. "I should say so; the whole division knows him. He can kick a football farther than any man I ever saw. He's a good soldier, too."

Brown's reputed civilian exploits included punting a ball on fourth down ninety yards over the goal posts at the opposite end of the field in a Jefferson City game.

Equipment Lacking

Rifles have been issued to the white troops, but the negroes have not yet been equipped. *(I will not make the obvious comment here.)*

"I saw them drilling the other day," said a white sergeant. "Armed with their wooden guns, they were stepping out as smartly and as smoothly as any troops you ever saw. The alignment was perfect."

Has there been any trouble over refusal of white privates who refuse to salute negro officers?

None, answered Major John C. Lee of the 89th Division.

"When a negro officer comes among white troops, he generally walks with his eyes straight ahead. Whether a white man salutes or not depends on what part of the country he is from. Race prejudice does not figure in the Army regulations, but it is human nature.

"Personally, I never fail to salute a negro who is entitled to the salute."

3 WAYS OF VIEWING COLORED SOLDIERS

February 16, 1919 *(printed in the Post-Dispatch with no byline, but a Washington dateline indicates it was written by Ross).* It is now possible to take stock of the conduct of our negro soldiers and report their notable work and heroic accomplishments.

Three ways can describe the colored man as a fighter in a war that called for more pure nerve than any other because of the terrible weapons used — to show him individually, then in a regiment engaged alongside white troops, and finally in a division composed of men of his own race.

As an Individual

The 368th Infantry, colored, fought in the Argonne. A runner had to be sent with a message to the left side of the American line, across an open field swept by heavy machine-gun fire.

Private Edward Saunders of Company 1 volunteered. Before he had gone far, a shell cut him down. As he fell, he cried to his comrades: "Come and get this message. I am wounded."

Lieutenant Robert L. Campbell ran across the shell-swept space, picked up the private, and with Germans hailing bullets all around him, carried his man back to the American line.

Both were awarded the Distinguished Service Cross, and Campbell was recommended for captaincy.

We also have learned how the same lieutenant outwitted the Germans.

Campbell knew by the direction of the Boche bullets that his party was unobserved, but he couldn't see where enemy fire was coming from. He ordered one of his men to crawl to a copse of thick underbrush and tie a rope to the stem of a bush, then to crawl away as far as possible and pull the rope to shake the brush.

The Germans poured forth their fire on the quivering brush. Now that they knew where the enemy was, the Americans dashed forward and killed four Boche with grenades, capturing three others.

With White Troops

The 372nd Regiment was the first negro group to go over, practically all its line officers being colored, sandwiched among the French forces.

They arrived in France on April 14 and went into training with the French on April 23. On June 6, the 372nd was sent to the trenches just went of Verdun, occupying the famous battle-swept Hill 304 and sections at Four de Paris and Vauquois.

On Hill 304, thousands of French and German soldiers had fallen; that this hill was given to the negroes to hold shows the confidence they had already won from the French.

The regiment's first engagement was in the Champagne sector, with Montoir as the objective. Here came the real test: To back up their enthusiasm, did they have the staying power drilled into European troops through centuries of training in the science of warfare?

In the nine days' battle, the negroes stood out with such distinction that the regiment won the coveted Croix de Guerre.

The 368th Infantry fought with the French 38th Corps in the Argonne Forest action as well.

All-Negro Division

Soon after the entire 92nd Division was thoroughly organized, it took over the Marbache section, where their activity in trench raids led the Germans to call them "Black Devils," but the division as a whole had never been in a set battle.

Their chance came in the drive on Metz. They were notified at 4 a.m. Sunday, November 3. The motto "See It Through" of the "Buffaloes," the motto of the 367th Infantry, echoed through the whole division.

They began their advance at 7 a.m. from Pont-a-Mousson. Before them was a valley commanded by the heavy guns of Metz and nests of German machine guns.

The negroes seemed to realize that here was their greatest opportunity to show their mettle — for the first time they would battle as a division. A sense of race solidarity possessed them, uniting their purpose as no amount of drilling could, and they were a terrible foe as they plunged forward to Preny.

So rapidly did they advance that the list of casualties, considering the rain of shells, was small. Their objective was Bois Frehaut.

Picked Moroccan and Senegalese troops of the French — in an odd competition of colored races — were the first to arrive. The Germans, seeing what was up, pounded Bois Frehaut with a heavy fire. It became too hot for the Africans; they were forced to retreat.

Were our colored fighters really going to "see it through"? The 56th Regiment was forced to withdraw, but not until it had stood up and borne a heavy loss. The First Battalion of the "Buffaloes" was called upon to hold the Germans at bay while the hard-hit 56th retreated. Here, in confronting the enemy with an iron resistance, the "Buffaloes" won their Croix de Guerre.

A little later, Bois Frehaut was taken by the 92nd. The Germans directed a murderous fire, but there was no driving out the colored men.

The 92nd suffered a total of 1,478 casualties, and numerous officers and enlisted men were commended for meritorious conduct.

Missouri-born Ross later became White House press secretary to his high-school friend, Harry S. Truman, the President who desegregated the U.S. armed forces in 1948.

• • •

When the black troops came home, they got a stirring, segregated, welcome.

A THOUSAND NEGROES ON PARADE

April 14, 1919. *(Post-Dispatch, no byline.)* Four hundred negro soldiers of the 92nd Division who saw active service in France and about six hundred National Army negroes who were trained in this country paraded through St. Louis streets today.

Those thousand soldiers were about the same number as the white troops who were honored with a march through downtown two weeks later and whose return was emblazoned with front-page headlines and photographs (Chapter 31). The return of the black troops took less than a column on Page 5 in the Post-Dispatch. The Star had no story.

The march was under the auspices of the St. Louis Boosters, an organization of negroes. A parade float held a figure representing the Goddess of Liberty and another held an American eagle. In the center a mounted figure depicted Joan of Arc.

A holiday was declared in negro schools for the occasion. Great crowds of negroes turned out, and each band had an admiring crowd of followers. Negro houses along the streets were decorated, and virtually every negro who could get off from work was there to see it.

The parade was cheered through the African-American district of the time: Along 4th, 12th, 14th, 19th, and 20th, Broadway, Channing, Chestnut, Lawton, Market, Morgan, Olive, Pine, and Washington.

Children and some adults were clad to represent races, including American Indians, Chinese, Japanese, and Hawaiians.

The Coliseum was engaged for the afternoon and night, and a "midnight ramble" later on. Lieutenant-Colonel Otis B. Duncan, who commanded a unit of the 369th Infantry, will speak, and the Boosters Club will present each man with a gold button to commemorate his service.

About the Author

GEORGE GARRIGUES HAS BEEN A reporter and editor with the *Los Angeles Times* and the head of journalism or communications programs at University of the Pacific, Wayne State University, University of Bridgeport, and Lincoln University of Missouri.

He has also worked on the *Inglewood Daily News* (California), *Ontario Daily Report* (California), *San Francisco Examiner, Coast-Valley Journal* (Oregon), Wave Newspapers (Los Angeles), and *Bergen County Record* (New Jersey).

To get more information and to contact him, go to *www. CityDeskPublishing.com.*

www.ingramcontent.com/pod-product-compliance
Lightning Source LLC
Chambersburg PA
CBHW051905090426
42811CB00003B/465